Nullius in Verba
Darwin's greatest secret

Volume 1

MIKE SUTTON

Vae Victus
Second edition and first print edition (vol. 1).
Abridged, updated, and with minor errata corrections.
First e-edition published by Thinker Books of Thinker Media Inc. USA 2014
In accordance with the U.S.A publisher's and author's international contractual agreements.

Front cover design by Andy Sutton of Vae Victus. Modified from the 1871 artwork
of James Tissot.

ISBN: 1541343964
ISBN-13: 978-1541343962

FOR ALL ORIGINATORS AND FIRST
DISCOVERERS

'... if the first description was originally imperfect, & had been superseded by any better description, it would perhaps be better to omit all reference to it, for the sooner such an author's name was buried in oblivion the better.'
(Charles Darwin 1849)

'This discovery recently published by Mr. Darwin turns out to be what I published very fully... as far back as January 1, 1831... reviewed in numerous periodicals, so as to have full publicity...by Loudon...and repeatedly in the United Service Magazine for 1831 etc.'
(Patrick Matthew 1860)

'I think that no one will feel surprised that neither I, nor apparently any other naturalist, had heard of Mr Matthew's views...'
(Charles Darwin 1860a)

'I notice in your Number of April 21 Mr. Darwin's letter honourably acknowledging my prior claim relative to the origin of species. I have not the least doubt that, in publishing his late work, he believed he was the first discoverer of this law of Nature. He is however wrong in thinking that no naturalist was aware of the previous discovery. I had occasion some 15 years ago to be conversing with a naturalist, a professor of a celebrated university, and he told me he had been reading my work "Naval Timber," but that he could not bring such views before his class or uphold them publicly from fear of the cutty-stool, a sort of pillory punishment, not in the market-place and not devised for this offence, but generally practised a little more than half a century ago. It was at least in part this spirit of resistance to scientific doctrine that caused my work to be voted unfit for the public library of the fair city itself. The age was not ripe for such ideas, nor do I believe is the present one,..'
(Patrick Matthew 1860a, p.433)

'...an obscure writer on Forest Trees, in 1830, in Scotland, most expressly & clearly anticipated my views — though he put the case so briefly, that no single person ever noticed the scattered passages in his book...'
(Charles Darwin 1861a)

'Unfortunately the view was given by Mr. Matthew very briefly in scattered passages in an Appendix to a work on a different subject, so that it remained unnoticed until Mr. Matthew himself drew attention to it in The Gardeners' Chronicle, on April 7th, 1860.'
(Charles Darwin, 1861, p.xv)

'...I was also ambitious to take a fair place among scientific men,—whether more ambitious or less so than most of my fellow-workers, I can form no opinion.'
(Charles Darwin in Darwin, F. 1887 p.65)

FOREWORD
BY THE AUTHOR

I wrote this book for two main reasons. Firstly, because of the egregious failure of the scientific community to solve the problem of the easily proven dishonesty of Charles Darwin's audacious claim that before 1858 "no naturalist" (Darwin 1860a) and later "no single person" (Darwin 1861a) ever noticed Patrick Matthew's prior-published original conception of the theory of macroevolution by natural selection. And secondly, because in this case the scientific community has uniquely and unjustly ignored its own firmly established Arago Case conventions and norms for awarding first and foremost priority to agreed first published original discoverers.

The science problem of Darwin's and Wallace's audacious claims to have each independently conceived Matthew's prior published origination of the theory of macroevolution by natural selection is not simply a criminological case of original scientific discovery, institutional injustice and plagiarising science fraud by glory theft. In this story, we find important information on issues of industry, agriculture, empire, heritage, sustainability, privilege, power, mythology, dishonesty, lies, radical politics, religion, heresy, sedition, science hero cultism, disreputable behaviour by academics, punctured paradigm protectionism, cryptoamnesia, "ethics of memory", blind sight, outright fact denial, cognitive dissonance and canny indifference to the newly discovered facts.

The dedicated promotion and celebration of Darwin and Wallace, at the expense of Matthew, is a remarkable example of bias, because the publication of Matthew's (1831) origination of the process of macroevolution by natural selection is a plausible causal intermediary for Darwin's (1839) private Zoonomia notebook, his (1842) and (1844) private essays, Linnean Society paper (1858) and *Origin of Species* (1859). The same reasoning applies for Wallace's notebooks and his (1855) *Sarawak* and (1858) *Ternate* papers.

For my investigation of this strangely neglected case, I adapted the rational approach that Matthew, and subsequently Darwin, used to argue for organic evolution by descent. That adaptation involved analysing the correlated evolutionary "selection" processes in the publication record, rather than nature, to demonstrate how it is possible to explain great similarities of "type" between Darwin's (1859) *Origin* and Matthew's (1831) prior published, read, advertised and cited predecessor *On Naval Timber and Arboriculture.*

The independently verifiable original evidence I discovered in 2013, first published in *Nullius* in 2014, includes important new facts that prove, before Darwin or Wallace put so much as pen to private notepad on evolution, their associates, correspondents, influencers, influencer's influencers and facilitators read, then cited, and some commented upon, the original ideas in Matthew's (1831) book.

That discovery is independently verifiable, confirmatory evidence for plausible Matthewian influence by descent, and for Darwin and Wallace having committed plagiarising science fraud by glory theft. Yet, Darwin went to his grave fallaciously denying that anybody read it who could have plausibly influenced him or Wallace, with the original theory in Matthew's book.

Asimov (1983) wrote that he would believe anything so long as the evidence was adequate: '*The wilder and more ridiculous something is, however, the firmer and more solid the evidence will have to be*'. The trouble is "wild" and "ridiculous" are subjective concepts. It seems wildly ridiculous to me that anyone would believe, in light of what we now know about whom they knew and what those influencers of theirs knew about Matthew's book and the original ideas therein, that Darwin and Wallace probably, or definitely, independently conceived the theory of evolution by natural selection. On the other hand, despite that "New Data", what remains seemingly wild and ridiculous to many Darwin scholars is the notion that their adored namesake was a lying plagiariser who committed the world's greatest science fraud by glory theft. It follows, that for those Darwinists, perhaps only the new discovery of a smoking gun confession from Darwin or other pre Gardeners' Chronicle (Darwin 1860a) evidence, asserting he definitely knew what was in Matthew's book, will do. If so, whether such a thing is found or not, history will judge them accordingly. On which note, influential Darwin scholars should not fact-deny that the newly discovered facts in this book prove that Darwin's *nobody read it* fallacious excuse for not finding and reading Matthew's book is disproven. The proven fact-led rupturing of that excuse means it now cannot be used, as it has been in the past, as the sole premise to support Darwin's and Wallace's claims of dual independent discovery. To do that now would be just as wild and ridiculous as to assert that it seems unlikely anything new has been discovered in the story of Matthew, Darwin and Wallace.

The excuse, penned by Darwin, for not having read Matthew's work was created as a deliberate lie. We know that as a fact, because before Darwin constructed his *no naturalist / nobody whatsoever* read it defence, Matthew (1860) had already informed him of other naturalists who, before 1858, did read the original breakthrough in his book. Darwin even knew well the work of one of those naturalists, who Matthew had already named. He was the famous polymath John Claudius Loudon. Darwin owned copies of Loudon's famous books and other publications on trees and plants. He had heavily annotated those years earlier, and praised them in private correspondence. Darwin's private notebook of books and periodicals he read proves by his own hand that he had already read five publications that cited Matthew's (1831) book (The Atheneum 1839, Loudon 1831, Loudon, 1838, The Gardener's Magazine 1841, Memoirs of the Caledonian Horticultural Society 1814-1832). As we can see, Loudon wrote two of them. Darwin wrote down Loudon's name many times before 1858. Darwin knew perfectly well, therefore, who and what Loudon was. Although Loudon had been dead 17 years by 1860, Darwin surely knew he had been a fellow member of the naturalist Linnean Society. And Darwin did know Loudon was an influential naturalist, who influenced him, when he penned his great lie that no naturalist had read Matthew's theory before he replicated it.

With absolute certainty, since 2014, we now newly know that, as opposed to the old Darwinian claim of none (e.g. Darwin 1861, 1861a, de Beer 1962, and Mayr 1982), the Big Data discovered "New Data" proves that at least 24 people cited Matthew's (1831) book in the literature pre-1858. Seven were naturalists. Four of those seven were associates and/or correspondents of Darwin's and Wallace's influencers, their friends, associates, influencers and influencer's influencers. Additionally, three of the four were at the epicentre of influence of the pre-1858 work of Darwin and Wallace on macroevolution by natural selection. Also before 1858, Darwin (1847b, 1848) met and corresponded with one of them, a famous naturalist who Wallace (1845) claimed as his own greatest influencer.

At the time I am writing these words in the summer of 2017, the predominant discovery paradigm is one of tri-independent discovery of macroevolution by natural selection by Patrick Matthew, Charles Darwin and Alfred Wallace. Directly challenging it is the compendium of new fact-driven discoveries in this book that are powerful circumstantial evidence of plausibly probable pre-1858 Matthewian "knowledge contamination" of the brains of Darwin and Wallace, via their influencers and facilitators and their influencer's influencers and facilitators, and others, who we now newly know read and then cited Matthew's book and mentioned the original ideas in it in print. Moreover, this "New Data" proves they did so years before Darwin's and Wallace's supposedly independent replications.

Let us call this newly evidenced alternative for the current paradigm, the

"Probable Matthewian Knowledge Contamination Hypothesis" for Darwin's and Wallace's otherwise amazingly miraculous independent replications of a prior published and cited theory.

Following the "New Data" discovery, members of our institutions of science are currently at 'Ghandi Stage 1' of his four stages of majority view acceptance of veracity-driven change (de Grey 2009: p.91): *'First they ignore you, then they laugh at you, then they oppose you, then they say they were with you all along.'* Perhaps we will eventually reach Stage 4. We should, because rationally there can surely be only one originator and first influencer of other naturalists on macroevolution by natural selection. The facts prove his name is neither Charles Darwin nor Alfred Wallace. The facts prove absolutely that his name is Patrick Matthew.

Since publication in 2014 of the first edition of this book, Howard Minnick, who is Matthew's 3rd great grandson, has marshalled substantial evidence demonstrating Matthew's genealogical connections to several prominent and famous members of the British Admiralty, landed gentry aristocracy and royalty. Historian and resident of the Carse of Gowrie in Scotland, Peter Symon, discovered Matthew's long lost grave. Thanks to funding from the award winning Edinburgh Skeptics Society, I presented the "New Data" at the International Science Festival in Edinburgh. A similar paper was presented at the British Society of Criminology Conference in Liverpool, which was later peer reviewed and published as a scholarly criminology article in 2014. Another paper was given at the London Skeptics Society Sunday Lecture at Conway Hall. Talks were given at Skeptics in the Pub meetings in Sheffield and Teesside. I gave a public lecture on the "New Data" at the James Hutton Institute in Scotland, the basis of which I turned into a peer reviewed science article in 2015.

In 2017, the Carse of Gowrie Sustainability Group won British National Lottery Heritage funding to build the Patrick Matthew Trail and organise the Patrick Matthew Celebratory Weekend in Scotland.

In 2015, Nottingham artist Gabriel Woods painted the oil on canvas Blessed Virgin Darwin as a hilarious heuristic device and allegorical explanatory analogy for the original findings in this book. Gabe's painting pays homage to religious pictures of the Virgin Mary and child. You can view it on Patrickmatthew.com. Also in 2015, the visual artist Astrid Leeson unearthed a wonderful collection of long forgotten letters written by Matthew to Lord Kinnaird. They reveal the depth of Matthew's libertarian character and political reformism concern for the welfare of tenant farmers and education of local children.

In 2017, Nottingham Trent University Distinguished Professor of Psychology Mark Griffiths added further gravitas to the debate. We have co-written two articles on Matthew's discovery and priority, which Professor Griffiths will now submit to learned academic journals.

Memory studies alert us to the "ethics of memory" and influence. On which note, it would be remiss to fail to note that this book evolved from the 1967 Summer of Love, because the stuff that really helped me make the new discoveries in it stemmed from the San Francisco 1960's techno-hippie pioneers and their subversive ethos of free access to socially important data. Their radical ideas led to the birth of the personal computer, brilliantly intuitive software, Internet, World Wide Web and Google. All five high technologies facilitated the Big Data discoveries that completely disconfirm the excuses previously given and accepted for believing Charles Darwin and Alfred Wallace independently conceived Patrick Matthew's prior-published, radical and liberating theory of macroevolution by natural selection.

Big Data technology, the Darwin Industry never reckoned with that!

~~~

Research teaches us that independently verifiable embarrassing newly unearthed facts, such as the ones in this book, are likely to meet with fierce resistance from those supporting the ideas and associated claims they challenge and demolish (Kuhn 1962). Unsurprisingly therefore, since publication of the first edition of this book in 2014, the facts in it have been on the receiving end of dishonest fact-denial, context-blind rationalisations, attempted fact-suppression behaviour, dishonest misrepresentation of the "New Data", and other unscientific and unscholarly behaviour.

Remarkably, much of this unprofessional conduct has come from several university academics. Their actions and words are important data for historians and for the natural and social scientist alike. Accordingly, I have archived what they have written for future analysis, discussion and understanding.

The first edition of Nullius (Sutton 2014a) is a 601-page e-book published by Thinker Books of Thinker Media inc. This second edition is abridged, updated, revised and errata edited. Written for experts, and non-experts, *Nullius* is for those interested in learning the newly discovered and independently verifiable facts about the pre-1858 readership, proven, and probable, influence of Patrick Matthew's original and complete conception of macroevolution by natural selection. This abridged first volume is one of three print books evolving from the original. Volumes 2 and 3 are forthcoming, eventually.

Dr Mike Sutton (2017)

CONTENTS

ACKNOWLEDGEMENTS

Thanks are due to my colleagues, friends and family who listened patiently and offered opinions on the significance of the original research findings in this book. Special thanks are due to Phil Wane and Andy Sutton. Staff at the ethical PR agency Journalista kindly informed the Scottish press about the New Data in 2014. The press office for Nottingham Trent University shared the news with the Chief science editor of the Telegraph newspaper. Consequently, published newsworthy reports of this important book are in several national, regional and local British newspapers between 2014 and 2017.

Three excellent books on Patrick Matthew, written by the surgeon and human transplant scientist William James Dempster, enabled me to appreciate the significance of my original discoveries of Patrick Matthew's pre-1858 influence on naturalists known to have influenced and facilitated the work of Darwin and Wallace.

Donald R. Forsdyke, professor of biochemistry in the Department of Biomedical and Molecular Sciences at Queen's University Canada, generously offered to read my manuscript for the first edition. He sent me detailed insights in April 2014. I took on board some of his recommendations, but only as I interpreted them and saw fit. If there are any errors or omissions in the first and subsequent editions of this book, they are my own.

Mike Sutton 2017

1 INTRODUCTION

Facts are friends, not enemies of the truth. All verifiable facts are scientific, whether people like them or not. This book uses newly discovered and independently verifiable facts, which I first shared with the world in the first edition of this book (Sutton 2014a), to solve the science problem of Charles Darwin's and Alfred Wallace's (1858) audacious claims of dual independent conceptions of Patrick Matthew's (1831) prior-published unifying theory of biology. Facts solve the problem of Darwin uniquely replicating by shuffling Matthew's same four words "the natural process of selection" to uniquely re-name it "the process of natural selection". These facts solve the problem of Darwin replicating Matthew's exact same nursery versus forest trees unique analogy of differences to explain evolution by natural selection. These facts solve the problem of Darwin's and Wallace's replication of many of Matthew's other original and highly idiosyncratic examples to evidence his bombshell breakthrough. These facts solve the problem of why Darwin lied about what he knew about the structure of Matthew's (1831) book. And these facts solve the problem of why Darwin lied repeatedly about the influential naturalist he knew and another whose name he didn't, who read and commented upon the significance, originality and heresy of Matthew's breakthrough, long before Darwin and Wallace replicated it.

In 1858, Darwin and Wallace replicated Matthew's (1831) prior published theory of macroevolution by natural selection, then each called it their own. After they were told it was not, and admitted as much (Darwin 1860a, 1861a, 1861, Wallace 1879), they continued to call it their own.

Darwin and Wallace's proprietary behaviour and foremost claims, to what each admitted was a prior published complete first conception by Matthew, are undoubtedly without precedent or parallel in the history of scientific discovery. Furthermore, established acceptance of the veracity and validity of their appropriation is equally unique. In light of the facts, this is a very strange case indeed. What adds to that strangeness is the failure of our scientific community to investigate properly this discovery anomaly.

Before the first edition of *Nullius* (Sutton 2014a) no one had accurately investigated, with reference to the contemporary literature and by way of systematic historic evidence gathering, who read Matthew's breakthrough before Darwin's and Wallace's replications, and the peculiar behaviour and questionable honesty of Darwin and Wallace. Consequently, there remains a pervasive history of science myth that Darwin originated the theory of evolution by natural selection (e.g. Desmond, Moore and Browne 2007). Scientists, however, including the world's leading experts on evolution, have fully explained many times in esteemed publications, that Patrick Matthew (1831) alone originated the full concept. He called it a theory, presented it as a hypothesis in need of further testing, and went into published print with it 27 years before Darwin and Wallace. To learn more about what has been written specifically on Matthew's publication priority for conceiving what is properly called macroevolution by natural selection, see: Darwin (1860a, 1861a, 1861), Wallace (1879), Mayr (1982), Dempster (1996), Cock and Forsdyke (2008), Wainwright (2008), Dawkins (2010), Rampino (2011) Ford (2011) Sutton (2015) and Weale (2015).

Regardless of what you might read about the ancient Greeks and Romans, Erasmus Darwin, Curvier, Lamarck, Hutton, Wells, Blyth, von Humboldt, and any number of others who wrote just a basic bit about the ancient development hypothesis of organic evolution, experts will tell you the fact that none other than Matthew himself originated his own original breakthrough.

Appendix 1 of this book contains everything Matthew wrote in his book that is even remotely relevant to his theory of natural selection.

Matthew's (1831) theory is not stated and contained in just one clear paragraph of his book "On Naval Timber and Arboriculture" (hereafter *NTA*). His great breakthrough used natural science to provide an alternative explanation to divine providential intervention for the creation of each species. I am sure experts in the area can improve upon my interpretation, but effectively, his original theory is as follows:

Just as humans artificially select flora and fauna to suit their requirements, so does nature, in its predominantly steady state, select as a natural process only the most environmentally circumstance suited individuals. This natural process of selection explains the emergence and extinction of species, both between and after catastrophic geological and meteorological events. This means that species are mutable. This mutability of life forms happens in nature, over almost unimaginable lengths of time, when the number of naturally selected successful individuals within a population become of such a different variety that they become so distinct from their ancestors that they finally branch away from them to become a distinct species. In this way, different species came from common ancestors. By means of this process only do new species come into being, defined as such because they are incapable of breeding with either their immediate ancestors or significantly different others of common descent. Life forms inheriting traits that equip them with the best circumstance suited power of occupancy establish ecological niches, which repel invaders and ensure the survival of their currently best circumstance suited progeny. But all life forms may, as new arrivals, overtop those less able to compete in the ever changing environmental struggle and competition for existence.

Adapted and extended from Zirkle's (1941) excellent explanation of what he believed was Darwin's own theory, I can explain Matthew's (1831) original complex theory of the "natural process of selection" in ten key points:

1. All individual organisms vary and so no two are exactly alike.
2. More individuals are born than can exist in the available space.
3. Consequently, there is a keen competition between all forms of life (struggle for existence) within their environment.
4. Permanence is an illusion. Things are always in some state of flux. Everything changes, evolves, over time. Occasionally catastrophic geological and metrological events occur. Species and individual organisms best able to survive changes between successive generations will outbreed those

less suited to particular changes of any kind.

5. The weakest or the least fit are eliminated in that struggle.

6. The fit survive.

7. Fitness to survive means being best circumstance suited to the environment. In rapidly changed and in relatively stable environments, the 'natural process of selection', ensures the fit survive and that fitness evolves, essentially as a *process* of *selection,* leading to changes to a significant percentage of members within a population of species.

8. The survivors form a basis for new variations.

9. An environment unfavourable to some varieties and species, but favourable to others, will eliminate some variations and entire species and select others.

10 These variations, selected over great lengths of time by the 'natural process of selection', produce dominant varieties, which eventually produce various new species by way of divergence from ancestors, from whom they are so different they can no longer produce fertile offspring together.

~~~

By way of an amusing explanatory analogy to demonstrate the unscientific weirdness of the Matthew case, according to Christian mythology, the Blessed Virgin Mary claimed she conceived Jesus of Nazareth independently of any other human. If she really did, then she did so whilst surrounded by men whose testicles were probably fertile, albeit to some unknown degree. Conceptually, therefore, that would be a supernatural miracle, which is why many Christians, particularly Catholics, believe it is. Fair play to them says I, even though I am an atheist, because they are admitting that they have a weird and irrational supernatural belief. Contrastingly, Charles Darwin and Alfred Wallace each effectively claimed in scientific publications to have conceived, independently of its originator Patrick Matthew, and of each other, the entire theory of macroevolution by natural conception. However, according to newly discovered data (Sutton 2014a) found in 19th century publications, both conceived it following direct and indirect influence in their field from men, whose brains were already fertile with it, albeit to some

4

unknown degree. Now, whilst we have no proof the testicles of those surrounding St Mary were fertile, we do have new proof those brains surrounding Darwin and Wallace were fertile with Matthew's ideas, and to whom some of those brains belonged, because their owners cited Matthew's (1831) book, which contains his breakthrough. Some even drew specific attention to Matthew's original ideas before Darwin or Wallace wrote a word on the same topic. Under circumstances such as that, if Darwin and Wallace conceived the theory of macroevolution by natural selection independently of any Matthewian knowledge contamination, from those they knew who read the original ideas in Matthew's book, would that make their conceptions analogous to a twin miracle?

Matthew (1831) referred to his theory as the "natural process of selection" and Darwin (1859) would later four word shuffle that apparently original phrase into his own apparently original version of it: the "process of natural selection"; even though he claimed never to have seen Matthew's original until Matthew (1860) brought it to his attention.

This supposed naming coincidence is, arguably, as incredible as the supposed additional amazing dual coincidence that Darwin and Wallace replicated Matthew's complex theory independently of its prior publication, even though they were influenced and facilitated by others who had already read it.

Matthew's original analogy of differences between artificial selection in plants and animals bred by humans, compared with those naturally selected in the wild, is the essential explanatory ingredient of his breakthrough. The fact natural "selection" is necessarily a "process" (see Howard 1982, p. 21) and the fact that this process is "natural" is essential to understanding the theory. That makes it a likely reason for why Matthew used all three words in his term, which is confirmed by the uncomfortable fact Darwin used the exact same three words in his four-word-shuffled version of it 27 years later.

Matthew's original analogy of differences is actually part of his theory of evolution by natural selection, because "selection" is related to adaptation. This is why Matthew (1831 p.76) used his original analogy of differences between artificial and natural selection to explain it, and why both Darwin (1844 and 1859) and Wallace (in Darwin and Wallace 1858) were compelled to replicate that as well, in order for their replications to make sense. So crucial is Matthew's analogy for explaining the theory that Darwin used it to open the first chapter of the *Origin of Species*, before replicating many other unique Matthewian explanatory examples (see Sutton 2014a for the fully detailed original analysis).

If all this essential replicating by Darwin and Wallace was incidentally merely coincidental and independent of influence from Matthew's original work, just how many other such multiple coincidences are required to render that explanation improbable, or else amusingly supernaturally miraculous?

Most tellingly, the esteemed anthropologist Loren Eiseley (1979) wrote that it could not reasonably be deemed coincidental, as opposed to evidence of his later plagiarism, that Darwin (1844) replicated Matthew's (1831) forestry expert and so exceedingly idiosyncratic timber and arboriculture analogy of trees artificially selected in nurseries versus those naturally selected in the wild.

We humans thoughtfully select traits in our breeding of plants and animals that are not necessarily suited to any organism's well being, only to our own want or needs. Analogously, as Matthew uniquely explained, unthinking nature selects, not purposefully like humans, but as an analogously different process, resulting in survival of the most circumstance suited, meaning the most "fit" for its particular circumstances in "the wild".

Matthew quite correctly allowed for geological and meteorological catastrophes in his model, but Darwin and Wallace never. Matthew's original theory of macroevolution by natural selection is, in every other relevant way, apart from that great superiority, virtually the same as Darwin's and Wallace's later versions. I believe, therefore, given the evidence of the unparalleled multiple replication of Matthew's original conception, words he used to name it, and his explanatory analogy of differences to explain it, that the question of Matthew's influence on them and the possibility that Darwin and Wallace committed science fraud by plagiarising glory theft is worthy of thorough investigation and objective explanation. I think that arguing against so doing, in light of these facts, is unscientific. If that argument is sound, then inquisitive and objective scientists and historians with a sense of justice and concern for veracity, rather than science saint worship, would agree. They would, wouldn't they?

The underlying scheme of Matthew's original theory is that all life, both animal and plant, is descended from just one originating primordial form that somehow became "alive". According to this theory of macroevolution by natural selection, there can be only one such originator. Analogously, according to the verifiable history of publication and its influence, Matthew was the theory originator, who combined existing ideas to create the bigger idea of macroevolution by natural selection receptive to his breath of life of unique hypothesis and theory. Darwin, therefore, is not capable of having been the natural father of the theory, Matthew is. Darwin's wife Emma wrote to Matthew (Darwin 1863) confessing the fact: '*He is more faithful to your own original child than you are yourself.*' Weirdly, like many facts in this book, you will not find that one in any book or learned article on Darwin; at least not in any written before this one. Similarly, I am not aware of any discussion of Matthew's religious heresy in any book or article about Darwin. But I have found a most telling discussion by a Darwinist of how originators view their scientific hypotheses (Howard 1982: p.95). Bizarrely, like so many books on Darwin, Howard's entire book makes no mention of Matthew. Instead he glorifies what the scientific establishment has entrenched as the acceptable story of 'Wallace's self-denial in the face of Darwin's priority…' He writes: 'As any scientist knows, a hypothesis is as precious to its discoverer as a child to its parents.' Perhaps Howard failed to mention the skeleton at Darwin's feast fact that Matthew's hypothesis was kidnapped by Darwin. The truth of Matthew's story should not be twisted into Darwin glorifying mythology, because he effectively claimed very stridently for many years (e.g. Matthew 1860, 1860a, 1860b, 1867, 1871) that it had been adopted without his prior consent and without due recognition of his parental rights to foremost as well as first priority. With the usual degree of blind sight weirdness, which so typifies biased Darwin worship, one scholar did try to twist that fact-led history into a new myth by de-facto fact-denial of all Matthew's post 1860 claims to his precious "child". As confirmatory evidence for the Dysology hypothesis, Professor Stott (2012. p.12) penned the following incredible new falsehood about Matthew: '*Flattered and mollified the fruit farmer published his final word on the matter in the Gardener's Chronicle on the 12th May.*' By that, Stott means his letter of 12th May 1860. In reality, as the facts I have just presented prove, not only was that far from Matthew's last word on his priority, that was the 1860 letter in which Matthew informed Darwin in no uncertain terms that he was wrong to write that no naturalist had read his breakthrough and understood it before Darwin adopted it as his own child!

Unlike Darwin (1861) and Chambers (1844), who each kept a creator god in their replica evolutionary models as the divine being who set things up, Matthew (1831), in Note F of his Appendix, heretically and uniquely mocked any such unevidenced belief in a miraculous divine supernatural originator of species. He did so with reference to natural science as a means of explanation, using what he referred to as a theory:

A particular conformity, each after its own kind, when in a state of nature, termed species, no doubt exists to a considerable degree. This conformity has existed during the last forty centuries. Geologists discover a like particular conformity - fossil species - through the deep deposition of each great epoch, but they also discover an almost complete difference to exist between the species or stamp of life, of one epoch from that of every other. We are therefore led to admit either of a repeated miraculous creation; or of a power of change, under a change of circumstances, to belong to living organized matter, or rather to the congeries of inferior life, which appears to form superior. The derangements and changes in organized existence, induced by a change of circumstance from the interference of man, affording us proof of the plastic quality of superior life, and the likelihood that circumstances have been very different in the different epochs, though steady in each tend strongly to heighten the probability of the latter theory.'

Darwin believed in his geological mentor's, Charles Lyell's, strict uniformitarianism notion that merely observing present geological conditions best explains the past and so there had been no great geological catastrophic events in earlier periods. Therefore, he simply rejected as Biblical mythology any notion of a geological or meteorological extinction event. Consequently, he (Darwin 1861) depicted the originator of the theory, which he replicated, as a 'catastrophist', thereby strongly implying Matthew credulously believed, unscientifically, in the Noah's flood biblical story.

Modern science, however, proves Matthew was right about species extinction events and Darwin wrong (Rampino 2011). Yet, before scientific understandings of such things, from the third edition of the *Origin* to the last, Darwin (1861) successfully portrayed Matthew as being less scientific than he was. Meanwhile, as said, at various times, it seems Darwin dared not reject the supernatural 'Creator' teachings of the Christian Bible. We can see an example of this unscientific behaviour in the third edition of the *Origin* (Darwin

1861, p.525):

There is a grandeur in this view of life, with its several powers, having been originally breathed by the Creator into a few forms or into one; and that, whilst this planet has gone cycling on according to the fixed law of gravity, from so simple a beginning endless forms most beautiful and most wonderful have been and are being evolved.'

As a man of science, years before either Darwin or Wallace first touched pen to private notepad on the same topic, Matthew combined his forestry, orchardist and arboricultural observations with existing published knowledge to solve the problem of the origin and extinction of species. Almost two entire centuries ago, he invited readers to test his breakthrough by further observations and experiment. The proof of this is in *NTA*. See Appendix 1 of the book you are currently reading for all his remotely and highly relevant original text on the topic.

Considering the "New Data" in order to weigh the evidence that, pre-1858, Darwin and Wallace were most probably influenced by *NTA*, it is important to remember that Matthew's book was not written in a language foreign to them, or published in some far away country. Instead, it is in English, was published in Darwin and Wallace's adult lifetimes on the same small island of their birth and longest residence. Then it was read, cited and prominently advertised by many of Darwin's London neighbours (see Chapter 3) in London publications, before he replicated the original theory and the highly idiosyncratic original examples that Matthew used to explain it.

NTA is about trees, oak trees, pine tree, apples, apple trees, other trees and much more besides. Now, there is an old saying: "the apple doesn't fall far from the tree". Let us consider for a moment what that really means and why we mean it when we say it, jokingly, literally and by analogy. Then consider the important fact that, most ironically, Darwin's (1859) work emphasised the importance of the evidence for descent with modification from the close geographic proximity of like with like in the geological strata, and also with reference to fossil location and the location of bones of extinct species and those living. As Howard (1982, p.43) puts it: *'The more recent a common ancestor, the closer its divergent products should be to each other in space.'* Before 1858, Darwin and Wallace, during their time in England, more so in London, were certainly concomitant in physical space with *NTA,* and with those who read and then cited it in their close physical proximity. Analogously, as close physical proximity of extinct ancestors is disconfirming evidence for divine creation, *NTA's* physical space presence, combined with the "New Data" on the physical proximity of its readership to Darwin, Wallace, their influencers and facilitators, is confirmatory evidence that Darwin's and Wallace's replications of the breakthrough in *NTA* descended from it, rather than a twin immaculate conception miracle.

There is lesson here about Darwin needing to take a leaf out of his own book regarding how he marshalled evidence such as that to argue, but not absolutely prove, one thing evolves from another. The telling question is why he, Wallace and others scientists failed to investigate the comparable and obviously similarly rational probability that the instigator of Darwin's "child" resided nearby in *NTA*.

Before publication in 1859 of Darwin's *Origin*, he and Wallace had been prior corresponding with one another (e.g. Darwin, 1848, 1857b), and meeting with each other's influencers (see Bates and Wallace 1848, Lyell 1881), men such as Robert Chambers, Charles Lyell and William Hooker, who was father of Darwin's best friend Joseph Hooker.

Arguably, in light of the facts about who he knew who knew about *NTA* at the time, Darwin's claim of independent conception of Matthew's originality is of equal incredibility as Wallace's to have conceived Matthew's prior-published breakthrough independently in a supposedly brilliant eureka moment, whilst in an immediate state of exhausted recovery from a cognitively debilitating state of malarial delirium (e.g. see van Wyhe 2013). Thinking about that feverish story, Wallace's eureka moment claim is, arguably, even more improbable, since we newly know that Selby (1842), chief editor of the journal that published his Sarawak paper (Wallace 1855), had earlier cited *NTA* and mentioned the original and then heretical ideas in it about certain trees growing better outside their "natural" habitats.

Darwin (1847b) corresponded with Chambers. We now know, that Chambers (1832) had earlier read and cited Matthew's book, before anonymously authoring his own heretical book (Chambers 1844), which put evolution in the air (Millhauser 1959) in the first half of 19th century. This is the same Chambers, who met and corresponded with Darwin (e.g. Darwin 1847b) a decade before Darwin's and Wallace's (1858) replications of Matthew's (1831) original prior-published breakthrough.

All these newly discovered independently verifiable facts disconfirm the current belief-based premise that no one who could plausibly have influenced Darwin or Wallace read Matthew's book before they replicated the original bombshell ideas in it. And there are many more such disconfirming facts to come in this book.

Now let us turn to consider the facts that disconfirm the validity of Darwin's and Wallace's effective claims to foremost priority for Matthew's theory. Let me explain. And it really is this simple: because Matthew got there first, and because Darwin and Wallace admitted that his *NTA* was a complete prior-published articulation of what they wrote on the same discovery, and even more surely because so many leading biologists agree that he did conceive it first, Matthew has both absolute first and absolute foremost scientific priority for it. At least he does according to conventions and norms on scientific priority, as classically described by Merton (1957).

In light of the facts, why have so few people heard of Matthew? Why is Darwin's face instead of Matthew's face currently on the back of the British £10 note? Why does Wallace's head adorn a Royal Mail postage stamp, but not Matthew's? Why do an imposing statue of Darwin and a portrait of Wallace adorn the grand entrance hall of the Natural History Museum in London, where there is nothing there currently to celebrate Matthew? Why did Darwin refer to natural selection as "my theory" on 43 pages of the *Origin of Species* (1859), and then, after being compelled to admit it was untrue (Darwin 1860a), continue to do so in every succeeding edition of that book?

The truth is that macroevolution by natural selection is no more Darwin's theory than it is yours or mine. Why then does everyone think of it as Darwin's theory? What happened to Patrick Matthew's rights to recognition, not only as first but also foremost originator?

Let us face it, years before 1858, Matthew's breakthrough was read and cited by Darwin and Wallace's influencers, facilitators, influencer's influencers and facilitators. Before my original discovery in 2013 of that "New Data" (see Sutton 2014a), Dawkins (2010, p.206) wrote, within the limits of what he knew then, a most revealing paragraph on the Matthew priority problem:

'I singled out Darwin and Wallace as the two nineteenth-century naturalists who independently solved the riddle of life. But claims of priority have been made on behalf of at least two other nineteenth-century writers, Patrick Matthew and Edward Blyth. If those claims are upheld, it should be a matter of some national

pride that all four independent discoverers of natural selection were British.'

In those few lines, Dawkins reveals his "knowledge belief" that neither Darwin nor Wallace read Matthew's book. The same premise has it that neither did they hear nor read as much as an influential whisper or fragment of its intellectual breakthrough bombshell contents. Weirdly, Dawkins reveals also that he is not prepared to accept evolution by natural selection as Matthew's theory, despite the fact he admits elsewhere in his chapter that only Matthew got the entre thing first. I immediately realised on reading Dawkins' account, that by arguably betraying every scientific convention of priority, without once mentioning them, he celebrates others for simply being more famous after they replicated Matthew's prior-published and prior-cited ideas and then failed to cite him. Having done that, Dawkins goes on to celebrate three naturalists besides Matthew simply because he believes all were "independent discoverers", all supposedly having failed to read or otherwise learn of Matthew's prior-published discovery. That strikes me as weirdly unjust, deliberately biased, or else blinkered and credulous lack of thinking.

Notably, Dawkins is completely wrong in my opinion to include Edward Blyth as any kind of candidate for the laurels of unchallengeable independent discoverer of evolution by natural selection. Firstly, Blyth, just like Darwin and Wallace, never wrote anything remotely relevant to natural selection until after Matthew's conception of it was published in 1831. Secondly, Blyth, unlike Matthew, Darwin and Wallace, never once claimed to have independently discovered Matthew's original discovery. Importantly, it is newly discovered (Sutton 2014a) that the editor and publisher of Blyth's (1835, 1836) influential papers on evolution was the famous naturalist and polymath John Loudon. That discovery is important because he reviewed *NTA* in 1832, and therein he remarked that Matthew had seemingly discovered something unique on the 'origin of species', which is the very title Darwin chose almost three decades later for his theory replicating book! Added to that additional, do you suppose "mere multiple coincidence", in our rapidly snowballing Pythonesque "mere multiple connected coincidences or not, as the case may be" collection, is the fact Darwin (1861) admitted Blyth's pre-1858 influence on his thinking, and met him years before 1858.

Now, following those facts is a further Howitzer bombardment of related evidence for probable Matthewian influence. Here it comes.

Before the publication of his *Ternate paper* (Darwin and Wallace 1858) on evolution by natural selection, and before Darwin's (1859) *Origin*, Wallace read and made notes on page 62 of his private *Species Notebook* about Blyth's (1855) work on the evolution of varieties (see Costa 2014). Furthermore, Blyth and Lyell then read Wallace's subsequent (1855) Sarawak paper and shared their ideas on its great importance with Darwin pre-1859 (see Beddall 1972). All of this particular, shall we suppose "merely multiply multiplying coincidental evidence", amounts to the discovery of a now proven route for oral or written original Matthewian knowledge contamination influence (Sutton 2015) between Loudon, Darwin and Wallace, via Blyth. Although that does not prove the route was taken.

In a science article (Sutton 2015) on the topic of these newly discovered routes for such evolution of the *Vestiges*, *Origin*, *Sarawak* and *Ternate* papers by descent with modification from *NTA*, I propose a threefold guilt-escalating typology of the ways "knowledge contamination" can occur and, in light of the New Data, could well have occurred in the cases of Darwin's and Wallace's replications.

1. ***Innocent Knowledge Contamination***: (a) Reach of original ideas in a prior-publication via subsequent published sources on the topic, which failed to cite the Originator as their source. (b) Word of mouth and/or correspondence to the replicator by those who read the Originator's work or communicated with others on the topic, who did understand its importance in whole or simply in part, but failed to tell the replicator about the prior-publication's existence.

2. ***Reckless or Negligent Knowledge Contamination***: (a) The replicator reads the original publication, absorbs information such as original ideas, examples and terms, but forgets having read it, and never does remember. (b) The replicator reads the original publication and takes notes, but forgets the source of the notes. (c) The replicator is told about original ideas by someone who understands their importance completely or simply in part, who explains they come from a publication, but the replicator fails to ask the name of the author and title of that publication.

3. ***Deliberate Knowledge Contamination*** (science fraud): The replicator reads the original publication, or is told about its contents and existence, takes notes, or is given notes, remembers this, but pretends otherwise.

To repeat the point already made, to be even-handed, the newly discovered increased, albeit arithmetically unquantifiable, reasonable likelihood that Matthew's conception did influence Darwin's and Wallace's replications of it is not proof it probably happened. This is because we cannot yet, and maybe we never will, know how to calculate mathematically such an increased probability at the individual level. Likewise, the current absence of any "smoking gun" written proof that it did happen is not proof that it probably did not.

In all reasonableness, in this particular case, it is perhaps rational to claim that absence of evidence it happened is not evidence of absence of the possibility it did. This is because many of the original pages were removed and now remain missing from Darwin's private notebooks. Furthermore, and importantly, we must not forget that many of Darwin's and Wallace's pre-1860 letters are known to be missing (Beddall 1968). Finally, even disregarding for a moment all the torn out pages in his private notebooks, any claim that Darwin's private notebooks or essays definitely prove he conceived the idea of natural selection independently of Matthew or Wallace is unsubstantiated by evidence. This is because there is no evidence that the dates he wrote on any of them in the privacy of his own home are accurate. No one confirmed any of those dates before 1858 (see Sutton 2015). Arguably, therefore, we cannot any longer give Darwin the benefit of the doubt in this regard or any other, because, as the book you are currently reading goes on to prove, he very clearly repeatedly penned falsehoods about what he actually knew of Matthew's influence on other influential naturalists years before he replicated Matthew's conception. In sum, independently verifiable evidence in this book reveals that since he wrote dishonestly on what he actually knew about Matthew's pre-1858 readership and influence, the point of his own pen punctures the myth of Darwin's legendary honesty.

Importantly, on the grounds of Darwin's now newly proven lies in this area, the historic usefulness of another accepted "Darwin excuse" is cast into doubt, because by dint of his proven lies he has now, arguably, lost any objective or even subjective right to the benefit of doubt. The point being, Darwin's first response claim in the Gardeners' Chronicle in 1860, that he had to purchase a copy of *NTA,* to see what was in it in order to reply to Matthew's (1860) letter, most certainly is not proof that the proven serial liar Darwin was telling the truth about not owning a copy at that time. To simply believe such a thing, in light of the independently verifiable facts about Darwin's downright dishonesty about Matthew and his book, would be unscientific, credulous and extremely biased to say the least. The fact of the matter is, that all we can know in this regard, is that Darwin may or may not have had a copy of *NTA* in his library when Matthew (1860a) first wrote to claim priority for his breakthrough.

Thanks to the IDD method that unearthed the New Data in 2013, this book proves that among the total of 24 people newly discovered to have cited it before 1858, at least seven naturalists did read the original ideas in *NTA* before 1858. That new knowledge is veracious, because those naturalists actually cited Matthew's book in the published literature.

The fact that three of the seven newly known naturalists who cited *NTA* pre-1858, Loudon (1832), Chambers (1844) and Selby (1842), were right at the epicentre of influence and facilitation of Darwin's and Wallace's published, so-called independent, discoveries of natural selection, is a unique and new discovery about the greatest scientific discovery ever made. I think it is one of the most important of the 21st century, because, as we have seen so far, it overturns every dishonest excuse invented by Darwin, and deployed by a multitude of unquestioning Darwinists (Sutton 2015), to rebuff Matthew's right to rightful recognition as one of the immortal greats of science.

Darwin (1861) admitted the influence of Chambers and Blyth on him and on others. Wallace (1845) wrote that Chambers' *Vestiges* was a great influence for his own endeavours to find evidence to solve the problem of species. By way of example, the newly discovered facts about the certain influence of Matthew on Darwin's and Wallace's influencers and influencer's influencers make a fool's errand of Bowler's 2013 book, *Darwin Deleted*. This is because Bowler, in effort to promote Darwin as both an original thinker and independent discoverer, seems to write that Wallace, unlike Darwin, never used artificial selection as an analogy to explain natural selection. In fact, Wallace does. Just like Matthew and Darwin, he used it (see Wallace in Darwin and Wallace 1858). And what are we to make of Desmond, Moore and Browne's (2007 p.1) claim that 'Charles Robert Darwin...' is the '...originator of the theory of natural selection...', when leading evolutionary biologists, including Darwin (1861), Wallace (1879a), de Beer (1962), Mayr (1982), Dawkins (2010),

Rampino (2011), and Weale (2015) wrote in no uncertain terms that Matthew is the originator?

How then should a scholarly book open on Matthew, if not to assert the fact that he, not Darwin, is the originator? Rationally, there cannot be two originators. The true originator's origination was published decades before Wallace's and Darwin's replications. So why are the editorial boards of prestigious publishing houses, such as Oxford University Press and the University of Chicago Press, publishing such weird claims that run contrary to the known facts?

Since 1859, Darwin scholars have been systematically parroting incredible irrational fallacies about Matthew, Darwin and Wallace. For over a century, such fallacy spreading has dominated the established academic literature on the history of discovery of the theory of macroevolution by natural selection. Perhaps we should expect this. After all, would it be unreasonable to expect those labelling themselves 'Darwinists', 'Darwinites' and 'Darwinians', who sit on the editorial boards of publishing houses, be capable of objectively assessing evidence for their namesake's lies and associated plagiarism by glory theft? Additionally, is any Darwin scholar likely to receive, for example, a prestigious Darwin Medal from the hugely influential Royal Society for absolutely 100 per cent proving that its members, past recipients of that medal, have disseminated credulously, in prestigious expert peer-reviewed publications, fallacies posing as truth, penned first by Darwin, about Matthew's prior-readership?

On which note, the Royal Society awarded the first three Darwin Medals to its namesake's close friends and scientific associates: Hooker, Huxley and Wallace. Let us consider what we know about them, to see just how revealing that fact is.

The first medal recipient, Hooker approved as truthful, re-dated and then sent on to the Gardeners' Chronicle, Darwin's (1860a) reply to Matthew's (1860) claim for priority (see Darwin 1860e). In that reply, Darwin wrote the falsehood that no naturalist had read Matthew's ideas. He penned that lie in response to Matthew's (1860) clear information that the eminent naturalist Loudon had reviewed his book. Then Darwin (1861) followed-up with another falsehood. This one being that Matthew's ideas were limited to his book's appendix, even though he wrote that falsehood in response to Matthew's (1860) published letter, which included large swathes of relevant text from the main body of *NTA*. Darwin's (1860e) letter to Hooker showed he thought it would be 'splitting hairs' to face the fact that Matthew's ideas were not limited to the appendix. Therefore, Darwin penned his *limited to its appendix* falsehood as another proven blatant lie. Following Darwin's (1860e) letter to him, Hooker was in no doubt, therefore, about that being a lie, he nevertheless colluded with Darwin's dishonesty because he approved it as being the honest truth by re-dating and sending Darwin's letter containing it to the Gardeners' Chronical, doing so in order to note his approval; exactly as Darwin requested.

Further independently verifiable evidence proves that the Darwin medal winner Hooker colluded with Darwin in a second lie. This is evidenced by the fact Hooker knew also that Loudon was a most famous and influential naturalist. We know this as a certainly because Hooker had earlier written most enthusiastically of the renowned botanist, Naturalist Magazine editor, fellow of the naturalist Zoological Society of London and of the naturalist Linnean Society, John Loudon's, superiority over other naturalists. That was almost two decades earlier, where he wrote (Hooker 1841, p.714):

'We should hardly do justice to our feelings, did we not introduce in our list of botanical publications, and did we not refer to a recent work of Mr Loudon's, as one of the highest importance and of the greatest utility to the arboriculturist; to every nobleman and gentleman of landed estate, who is desirous of improving his property, and enlarging the resources of his country; and to every botanist and cultivator who wishes to become acquainted with the trees and shrubs whether indigenous or exotic, which will bear the climate of Great Britain: we allude to the Arboretum et Fruticetum Britan-nicum or the Trees and Shrubs of Britain. In this ample and characteristic title, there is nothing promised that is not fully and skillfully performed; so skillfully that we will venture to say there is not a naturalist in Europe who could have executed the task with anything like the talent, and judgment, and accuracy, that are here displayed by Mr Loudon.'

In 1860, another famous naturalist, Robert Chambers, who we now know read and then cited Matthew's book in 1832, went on to write his own on arboriculture (Chambers and Chambers 1842). Rationally, it hardly seems like it could be, not a bombshell fact fragment, but simply another mere multiple coincidence that he would then anonymously author his own heretical book on evolution (Chambers 1844). Furthermore, what about the fact that four years later, Chambers met and corresponded with Darwin (Darwin 1847b,

1848)?

Fascinatingly, Chambers famously convinced the Darwin Medal winner Huxley to stay at the British Association for the Advancement of Science conference at Oxford in 1860b, by telling him not to desert Darwin's evolution by natural selection cause in a debate with Bishop Wilberforce (see Bowler 2007).

Almost two decades later, on September 4 1867, Darwin's associates Wallace, Lyell, and Chambers were present at the British Association for Advancement of Science annual meeting in Dundee. There, Matthew was scandalously platform blocked from speaking on his own discovery. Yet the British Association allowed others to hold forth in attributing it to Darwin (Matthew 1867).

As for the second Darwin Medal winner, Wallace, we must not forget that we now also newly know, and, to remind any Darwinist revisionist historians lest they choose to unethically forget, once again (e.g. see Telegraph 2014), this knowledge is only thanks to my original research (Sutton 2014a), that his *Sarawak* paper's editor was the naturalist Selby, who cited Matthew's book many times in (1842)!

Besides, Hooker, Wallace and Huxley, other Darwin medal winners include Mayr (1982) and de Beer (1962). The latter two credulously parroted as true Darwin's knowingly penned fallacy, easily discoverable as a lie at the time, that Matthew's ideas were unread by any naturalists before 1860. These facts, revealed so far, are, arguably, embarrassing at the institutional level for the Darwin honouring Royal Society, Linnean Society, and British Association for the Advancement of Science, as well as for a multitude of university departments, publication houses, individual scientists, historians, and a multitude of peddlers of overpriced shoddy Darwinian paraphernalia, collectively comprising the so-called (Howard 1982) "Darwin Industry".

Without Matthew's influence on those who influenced and facilitated the work of Darwin and Wallace, and on their influencer's influencers and facilitators, it is now newly possible to argue quite rationally that Darwin and Wallace might perhaps have published nothing of note, perhaps nothing at all, on evolution by natural selection.

To recap, thanks only to the technology of Big Data analysis, we now know Wallace was facilitated, and perhaps influenced by his editor Selby, who cited Matthew (Selby 1842) years earlier. Moreover,

Wallace admitted the great influence Chambers had on his work, and Big Data analysis uniquely revealed that Chambers cited Matthew in 1832. Moreover, to emphasise the point already made, Wallace's pre-1858 private notebook records he read one of Blyth's two articles that were edited and then published by Loudon (Costa 2014), who, it has been long known, cited Matthew in 1832. However, the fact Loudon was the editor of Blyth's (1835, 1836) articles is yet another brand new Big Data re-discovery of the 21st century.

Today, we know that compared with Darwin's replication of Matthew's theory, Matthew's original version is superior. The supremacy of Matthew's *'natural process of selection'* over Darwin's *'process of natural selection'* is supported by the adroit evidence-led arguments of eminent surgeon and evolution expert Jim Dempster (1996), as confirmed by the famous biologist Michael Rampino (2011).

What Darwin uniquely did that is so important was to gather from the literature many examples of confirmatory evidence for Matthew's original and prior-published theory. However, reason, justice and the Arago Case ruling has it that no amount of such confirmatory evidence gathering could ever transmute Matthew's prior-published theory into one of Darwin's original conception and foremost priority.

The discoveries of most originators and first proposers do not come out of thin air. Rather, they represent some kind of problem solving breakthrough arising from an original intellectual synthesis of existing knowledge, or else from new discoveries arising out of the observed outcome of experiments informed by prior-knowledge. The discovery of the theory of macroevolution by natural selection, before Darwin replicated it and successfully claimed it as his own theory in the *Origin*, is no exception.

The fact Darwin actually claimed in print to have had no prior knowledge of Matthew's prior published breakthrough, and claimed also that no other naturalist was aware of it before Matthew (1860) brought it to Darwin's attention in the press, is ignored by the famous Darwinist and founding director of the Skeptics Society, Michael Shermer. Instead, Shermer sidetracks us from this fact by telling us that most originators have their known influencers. Shermer uses this obvious truism to explain to us that we should not think, therefore, that there is anything at all unusual about Darwin's replication of Matthew's theory (Shermer 2002, p.147-149). He then tells us that the influence of any other thinker on Darwin is not, therefore, a zero sum question. However, that argument is, arguably, utter flim-flam, for the simple reason that Darwin claimed Matthew had zero influence on him and zero influence on any other naturalist. In other words, it was Darwin himself who claimed it was a zero sum game. He most surely did so simply to avoid evidence-led accusations of prior-influence and possible plagiarism. Why else would he do it? Most tellingly, it was a proven deliberate lie and, as we have seen, both Darwin and Hooker knew it and colluded in telling it.

Perhaps we should not be surprised to learn that in the same book Shermer (2002) turns his hand next to un-sceptically spread the myth started by Darwin that Matthew's origination was hidden away solely in an appendix of an obscure book and ignored. By way of presenting the disconfirming facts for such nonsense, Appendix 1 of the book you are currently reading contains all of Matthew's writing on natural selection from *NTA*, so that readers can see for themselves the facts of what Matthew wrote, how much he wrote and where he wrote it. The facts prove that Matthew's work on natural selection was not solely in the appendix of *NTA* at all. A great deal of it, including his name for it, his use of artificial selection as an analogy of differences to explain it, and his call for others to look for evidence to confirm it, is all in the main body of *NTA*. Small wonder Darwin (1861) invented the "Appendix Myth". We know that what flowed next from his own pen (see Darwin 1860e), reveals Darwin knew that appendix story was another crafty falsehood, one of his own devising, written, most surely, to steer others away from the facts about Matthew's breakthrough.

Is it wrong to call someone a liar if you cannot totally prove he or she meant to tell a lie? I think Princeton University Philosopher Harry Frankfurt (2005: pp.51-52) settles that matter:

Telling a lie is an act with a sharp focus. It is designed to insert a particular falsehood at a specific point in a set or system of beliefs, in order to avoid the consequences of having that point occupied by the truth. This requires a degree of craftsmanship, in which the teller of the lie submits to objective constraints imposed by what he takes to be the truth. The liar is inescapably concerned with truth-values. In order to invent a lie at all, he must think he knows what is true. And in order to invent an effective lie, he must design his falsehood under the guidance of that truth.'

If we accept Frankfurt's rational philosophical reasoning as valid, the verifiable evidence from his own pen proves Darwin was an artisan in dishonesty. He successfully succeeded in his aim to have the point of fact, of which Matthew twice informed him, of Matthew's influence on influential naturalists, occupied by a knowingly told falsehood. The falsehood of Darwin's devising being to write about something, which he knew happened, to claim it never happened. According to Frankfurt's expert knowledge about lies, that makes Darwin a liar. Moreover, and most importantly, that lie alone makes Darwin, at the very least, a plagiarizing science fraudster by post-hoc glory theft (see Sutton 2015) of the prior-published theoretical breakthrough he replicated and claimed as his own.

Since the age of the Great Enlightenment (Deutsch 2011) in the 18th century, when testable and potentially disconfirmable knowledge claims finally trumped the status of the claimant and their source, skeptical and independent analysis of claims has been a requirement of science (Potter and Wetherell 1987). Consequently, *Nullius in Verba,* the motto of the Royal Society, which is the oldest scientific academy in continuous existence, sums up what should have been the original approach taken to the problem of Darwin's and Wallace's claimed dual independent replications of Matthew's prior-published theory. Nevertheless, no one took it. Why on Earth not?

With no comparable cases of any kind of incredible dual replication event of a prior published theory and its unique explanatory examples, my further examination of newly discovered facts in the following chapters will hopefully help you weigh the evidence to reach your own conclusions about what to make of this weird case.

2 NULLIUS IN VERBA CHARLES DARWIN

This chapter revisits some of the information from Chapter 1 to examine in more detail what we know about who Darwin knew who definitely read *NTA* before his (Darwin 1859) and Wallace's (1858) replications and his proven audacious lies in that regard.

Following private and public criticism for failing to cite his influencers in 1859, Darwin began the third edition of the *Origin* with a chronological historical sketch of the precursory thinkers of organic evolution. Or, more precisely, it was an acknowledgement of those that he admitted shaped his thinking on the theory of natural selection and those who got there before him, whose work he claimed to have been unfamiliar with before 1859. The subject of Matthew's prior discovery of natural selection fell into the latter group. Darwin wrote (1861, p.xv):

'In 1831, Mr. Patrick Matthew published his work on 'Naval Timber and Arboriculture,' in which he gives precisely the same view on the origin of species as that (presently to be alluded to) propounded by Mr. Wallace and myself in the 'Linnean Journal,' and as that enlarged on in the present volume. Unfortunately the view was given by Mr. Matthew very briefly in scattered passages in an Appendix to a work on a different subject, so that it remained unnoticed until Mr. Matthew himself drew attention to it in the 'Gardener's Chronicle,' on April 7th, 1860.'

However, as I explained in Chapter 1, the best expert rational definition available (Frankfurt 2005) informs us that a lie is a falsehood told deliberately by its originator as the opposite to what they know to be true, in order to convince recipients of it that the falsehood is true. The facts prove Darwin (1861) wrote two such falsehoods in that paragraph of his historical sketch. They are two lies. As we know, one was that 'no naturalist' (Darwin 1860a), later 'no single person' (Darwin 1861a) read the original conception so that it 'remained unnoticed' (Darwin 1861).' The second is that Matthew's original ideas are contained solely in *NTA*'s appendix.

Dealing with the 'Matthew's original breakthrough was unread before 1860' lie first. Matthew's (1861) letter to the Gardeners' Chronicle informed Darwin that John Loudon (1832) had read and reviewed his book. Loudon was a famous naturalist. He was, for example, the owner and editor of the Magazine of Natural History (1832), which proclaimed on its title pages Loudon's prestigious membership of the Linnean Society, Zoological Society and various other naturalist societies in Europe. Although Loudon was dead by that time, Darwin knew Loudon's general work well. His notebook of books read (1838) contains reference to it and proves that he read five publications that cited Matthew's (1831) book (Sutton 2014a), two of them written by Loudon. Books from Darwin's library contain heavily annotated copies of Loudon's work on trees. Moreover, William Hooker, the father of Darwin's best friend Joseph Hooker, was a close friend of Loudon's friend and co-author John Lindley. The botanist Lindley wrote sections of Loudon's (1841) Encyclopedia of Plants, and Darwin (e.g. Darwin 1843) regularly corresponded with Lindley and contributed to the Gardeners' Chronicle during Lindley's chief editorship (Elliot 2010). Clearly then, Darwin would have been in no doubt whatsoever that Loudon had been an extremely important and well networked naturalist who was a member of naturalist societies, which one needed to be a noted naturalist to join.

So we know that Darwin knew that Loudon read Matthew's book and reviewed it, because Matthew (1860) told him so in the highly esteemed Gardeners' Chronicle. Yet Darwin (1860a) wrote in his reply to the Gardiners' Chronicle that no naturalist had read it.

To set the record straight with independently verifiable facts, once and for all, what follows is a simple chronological presentation of important and neglected historic facts, underlined here by me, for emphasis, which prove Darwin lied about the prior readership of Matthew's discovery and that Darwin's same lies have been credulously parroted as the truth ever since by the world's leading

Darwin scholars.

Patrick Matthew (1860) first open letter to Charles Darwin in the Gardeners' Chronicle and Agricultural Gazette 7 April 1860, pp.312-313, reveals that the famous naturalist botanist John Loudon reviewed Matthew's book:

This discovery recently published by Mr. Darwin turns out to be what I published very fully... as far back as January 1, 1831... reviewed in numerous periodicals, so as to have full publicity... <u>by Loudon</u> ... and repeatedly in the United Service Magazine for 1831 etc.'

Most notably, the famous naturalist Loudon (1832, p.703), who we now know was well known to Darwin and his closest friends and correspondents, wrote in his book review:

'One of the subjects discussed in this appendix is the puzzling one, of the <u>origin of species</u> and varieties; and if the author has hereon originated no original views (and of this we are far from certain), he has certainly exhibited his own in an original manner.'

Of equal notability, the United Services Journal and Military Magazine (1831) review of Matthew's book reveals that its anonymous reviewer also fully understood the admixture of seditious reformist libertarian politics and heresy of Matthew's bombshell breakthrough on the origin of species.

The United Services Journal and Military Magazine (1831) p.457:

'But we disclaim participation in his ruminations on the law of nature, or on the outrages committed upon reason and justice by our burthens of hereditary nobility, entailed property and insane enactments.'

Charles Darwin (1860a) replied to Patrick Matthew in the Gardeners' Chronicle and Agricultural Gazette 21 April 1860, no.16, pp.362-363:

'I think that no one will feel surprised that neither I, <u>nor apparently any other naturalist,</u> had heard of Mr Matthew's views...'

Patrick Matthew (1860), second open letter. Reply to Charles Darwin in the Gardeners' Chronicle and Agricultural Gazette 12 May 1860, p.433:

'He is however <u>wrong in thinking that no naturalist was aware</u> of the previous discovery. I had occasion some 15 years ago to be conversing with a <u>naturalist, a professor of a celebrated university,</u> and he told me he had been reading my work Naval Timber, but that he could not bring such views before his class or uphold them publicly from fear of the cutty-stool, a sort of pillory punishment... It was at least in part this spirit of resistance to scientific doctrine that caused <u>my work to be</u>

voted unfit for the public library of the fair city itself. The age was not ripe for such ideas…'

Charles Darwin (1861a), Letter to Quatrefages de Bréau, J. L. A. De. 25 April:

'…an obscure writer on Forest Trees, in 1830, in Scotland, most expressly & clearly anticipated my views — though he put the case so briefly, that no single person ever noticed the scattered passages in his book…'

And as noted above, Charles Darwin, (1861, p.xv) wrote in 'On the Origin of Species by Means of Natural Selection: Or the Preservation of Favoured Races in the Struggle for Life', 3rd ed.:

'Unfortunately the view was given by Mr. Matthew very briefly in scattered passages in an Appendix to a work on a different subject, so that it remained unnoticed until Mr. Matthew himself drew attention to it in The Gardeners' Chronicle, on April 7th, 1860'.

Gavin de Beer (1962), 'The Wilkins Lecture: The Origins of Darwin's Ideas on Evolution and Natural Selection', Proceedings of the Royal Society B: Biological Sciences. vol. 155, no.960, pp.321-338:

'…William Charles Wells and Patrick Matthew were predecessors who had actually published the principle of natural selection in obscure places where their works remained completely unnoticed until Darwin and Wallace reawakened interest in the subject.'

Ernst Mayr (1982), 'The Growth of Biological Thought: Diversity, Evolution and Inheritance.' Harvard University Press, Cambridge, Massachusetts, p.499:

'The person who has the soundest claim for priority in establishing a theory of evolution by natural selection is Patrick Matthew … His views on evolution… neither Darwin nor any other biologist had ever encountered them until Matthew bought forward his claims in an article in 1860 in The Gardeners' Chronicle.'

With reference to verifiable original sources so far cited in this book, it is proven that certainly not zero, and not just one, but at least seven naturalists read Matthew's theory before 1858. Furthermore, some of those were in Darwin's close social network. Consequently, Darwin was wrong, which proves that his oft-credulously parroted *no one read Matthew's original ideas before 1860 excuses* for his claim to have had no prior knowledge of Matthew's discovery are proven to be completely fallacious and outright plagiarizing science fraud, by post-hoc glory stealing, lies.

Putting the question of science fraud by glory-stealing lies aside just for a moment, the excuses given by the scientific community for denying Matthew's scientific priority and his greatness as the first published originator of, and first influencer on, natural selection are not only illicit, they are also totally and irrefutably incorrect. Moreover, they are based entirely on Darwin's proven lies.

Loudon (1832) reviewed *NTA* and highlighted the originality of Matthew's ideas on species and organic evolution. Weirdly, this is just one among many long known facts that many Darwinists have studiously ignored in their pseudo-scholarly efforts to replace the facts with fallacious nonsense substituting for good scholarship. All this, to thwart Matthew's claim to his own original breakthrough and its pre-1858 influence.

Much has been made by a few (e.g. Eiseley 1979, Davies 2008) of Darwin's cited acknowledgement of the influence upon him of some of Blyth's articles and his failure to cite others. Yet none, who note areas of Blyth's certain un-cited influence upon Darwin, take account of the fact that all of Blyth's ideas were published at least four years after Matthew's published breakthrough of 1831. More tellingly, his two key papers (Blyth 1835, 1836), which reveal an understanding of the natural process of selection, were published in the *Magazine of Natural History* in the time that it was both owned and edited by the Scottish botanist John Loudon. That fact, though previously ignored before the publication of *Nullius* 2014, is significant, because we know Loudon had earlier reviewed Matthew's book and remarked positively upon its author's unique ideas on what he referred to as 'the origin of species.' Consequently, the likelihood of Blyth's work benefiting from some form of Matthewian knowledge contamination via Loudon is clearly plausible. Therefore, both the post-*NTA* date of Blyth's published papers and his editor's prior-knowledge of Matthew's unique discoveries must, in absence of obvious bias, now engender doubt that Blyth's published work on the topic came independently of Matthew's unique breakthrough.

If Darwin's social connections tie-in with a considerable number of those who read Matthew's book, then it would be increasingly more likely than not that he would have learned of its existence during his 20 years of voracious research of the literature, and so read it himself, before Matthew brought it to his attention in 1860. On that reasoning, the formula devised for Darwin's innocence or guilt is a simple one. Namely, the existence of any number of natural scientists who read Matthew's book, whether connected socially to Darwin or not, disconfirms the validity of his excuse for being unaware of Matthew's theory within it. Moreover, the size of that number, connected socially to him, exponentially incriminates Darwin as a probable liar for saying that neither he nor any other naturalist known to him was aware of Matthew's ideas. The scientific community has to date simply taken Darwin's word for it when he claimed no prior knowledge of *NTA*, but such mere credulous belief is unacceptable. There is no rational excuse for continuing to grant Darwin, the replicator and proven liar, special privileges over the originator Matthew.

3 THE 24

My research uncovered 24 people who definitely read *NTA* before 1858. We know they read Matthew's book, because they cited it in the literature. I looked in detail at the lives of those people. I examined their interests and careers. Most importantly, I examined whether or not they were part of Darwin's close social network.

One reason why such a systematic approach has never before been undertaken with traditional methods is the enormity of the task. Without the new Big Data Internet Date Detection Method (IDD), which facilitated my original discoveries, where on the library shelves and archives should one start looking for undiscovered references to Matthew's work?

All the same, it is surprising; in light of the fact that even Darwin could not refute the reality of Matthew's complete prior published theory, that before my publication of *Nullius* none have tried to discover whether it was true that Matthew influenced no one who mattered in the field. Five telling questions follow: (1) Why, after Matthew (1860) told Darwin about it in the Gardiner's Chronicle, did nobody follow up the important Loudon lead? (2) Why, after Matthew (1860b) told Darwin about it, did none investigate who read and then approved banning his book at Perth public library? (3) Why, after Matthew (1860b) told Darwin about it, did neither Darwin, Wallace, nor any other scientist ask Matthew for the name of the unnamed naturalist, from an eminent university, who feared pillory punishment if he was to teach or otherwise disseminate Matthew's original ideas? (4) Why was Darwin so weirdly and unscientifically incurious about such major matters regarding what he called "my theory"? (5) Why was Wallace totally incurious about them as well?

In the 19th century, important facts provided by Matthew about his influence on other naturalists should have been followed-up at the time. Today, unless the historic letter, diary or undetected publication archives contain that information, the chance to answer these important questions is lost.

The hardest of literature to find is that which no expert in the relevant field has found. That I am no expert in the field of evolutionary studies, and yet I found so much previously undiscovered relevant information, is evidence of the revolutionary effect that Big Data IDD is likely to have in all areas of scholarship. The method really is as simple and effective at finding hidden treasures in the literature as a metal detector can locate a hoard of coins in the field of antiquity.

Before Darwin's publication of the *Origin,* the people that we know for sure read *NTA*, apart from Matthew and his publishers, are those who reviewed or cited it. The first known member of this group was the famous naturalist botanist, garden designer and editor John Claudius Loudon, who went on to write a highly favourable review of the book (Loudon 1832), where he highlighted the originality of Matthew's theory on the problem of species. Loudon then went on to cite *NTA* in several other important publications.

List 1 is a record of who we now know read *NTA*, because they actually cited the book pre-1858. This list includes anonymous authors. While it remains possible that one or more of these anonymous authors might be the same person, or that one of the named reviewers may elsewhere have anonymously reviewed the book, in the absence of any such evidence that they did so, it is presumed that they are different individuals, which seems, at least intuitively likely.

List 1

> Edinburgh publisher Adam Black
> London publisher Longman, Rees, Orme, Brown and Green (1831)
> *The Farmer's Journal* — Unknown reviewer (1831)
> *The Perthshire Courier* — Unknown reviewer (1831)
> *The Elgin Courier* — Unknown reviewer (1831)

The Country Times — Unknown reviewer (1831)
The United Service Journal and Naval and Military Magazine —
  unknown reviewer (1831)
The Edinburgh Literary Journal — unknown reviewer (1831)
The Metropolitan — unknown reviewer (1831)
John Claudius Loudon (1832)
Robert Chambers (1832)
John Murray II (1833)
John Murray III (1833) via the same publishing house as John
  Murray II
Edmund Murphy (1834)
Gavin Cree (1841)
John William Carleton (1841)
Cuthbert William Johnson (1842)
Prideaux John Selby (Selby 1842)
The Penny Magazine (1838) (1842) — Anonymous article
Publishers Cradock and Co. (1843) in "British Forest Trees"
Henry Stephens (1851)
John. P. Norton (1851) (Co-published with Stevens above)
Levi Woodbury (1832) (1833) (1852)
William Jameson (1853)
Wyatt Papworth (1858)

I would prefer to claim that the certain figure of those citing
Matthew is 24 and not 25, Firstly, this is because I think there is
sufficient reason to doubt that Carleton read *NTA*, since he merely
reproduced text within which Selby cites the book. It is important to
note that it is either Darwin's publisher, John Murray III, or his
father who cited *NTA*. This fact, combined with Murray III's
otherwise strange insistence that Darwin explain exactly from where
he got the term 'natural selection' (see Darwin 1859a), perhaps
suggests that Darwin's publisher was fully aware of Matthew's (1831)
distinctive phrase 'natural process of selection' and theory. Equally,
merely speculatively, maybe Murray III was concerned that Darwin
(1859) had, nine times in the *Origin*, originally, distinctively and
unimaginatively four-word shuffled Matthew's essential phrase to
'process of natural selection', in order to re-name the exact same and
prior-published breakthrough? However, we cannot know what
people knew about such things in the 19th century without further

evidence. We can merely wonder about it and look further for the existence, or proof of the lack of, such evidence in the archives.

As List 1 reveals, we now do absolutely know with 100 per cent certainty, contrary to the current myth, influentially disseminated by Richard Dawkins (2010), Michael Shermer (2002) and other famous and highly influential Darwinians of international standing (e.g. de Beer 1962, Mayr 1982), that many people did read Matthew's original natural selection ideas before 1858. We absolutely do know it, because we now newly know 24 people actually went so far as to cite *NTA* in the published literature before that year. What is on the printed page in the 19th century publication record of those citations definitely exists there. Although scientists do not like the idea of something being 100 per cent proven, being always found where discovered, the certainty of the existence of those printed and published words in the historic publication record is analogous to a natural law such as gravity. Anyone doubting that fact, who can float off and refute it with independently verifiable evidence, would make a fortune as an expert witness in libel trials, an immortal great discoverer and influencer in science and a small children's entertainer.

Most importantly, seven of those who cited *NTA* were naturalists. The seven, in date order of their citing of Matthew's book, are John Loudon (1832), Robert Chambers (1832), Edmund Murphy (1834), Cuthbert Johnson (1842), Prideaux John Selby (1842), John Norton (1851) and William Jameson (1853). Moreover, five of the seven: Loudon; Chambers; Johnson; Selby; and Jameson were part of Darwin's wider social circle. Loudon, Selby, Johnson and Chambers were in his inner circle, because they were directly linked to him through their mutual membership of scientific associations, including senior capacities at the British Association for the Advancement of Science, Royal Society of Edinburgh and the Linnean Society. Darwin's father had been Selby's houseguest. Johnson was Darwin's neighbour in Bromley. Selby, Johnson and Darwin were members of the Royal Society. Darwin corresponded with Johnson's brother. Significantly, he corresponded with, met and received as a present from Chambers a copy of his top-secretly authored, heretical work on evolution—*The Vestiges of Creation*.

As outlined in Chapter 1, these original findings represent a discovery paradigm-changing bombshell in the history of scientific discovery. This is because, contrary to the myth started as a lie by

Darwin (1860a, 1861, 1861a) other naturalists, including at least three known personally to Darwin and Wallace, had pre-1858 knowledge of Matthew's discovery. Moreover, those three naturalists were at the centre of Darwin's and Wallace's involvement in the field of macroevolution by natural selection.

The most influential papers of Blyth (1835, 1836), a naturalist who Darwin (1861) admitted had served him as a most valuable informant and influencer, were edited and published by John Loudon, who earlier had read and cited *NTA* (Loudon 1832), writing that Matthew had something original to say on "the origin of species".

Robert Chambers (1832), who both Darwin and Wallace freely admitted was a great influence on their work and on the minds of many others on evolution, read and cited *NTA* pre-1858. Reason suggests that reading Matthew's prior discovery of evolution by natural selection surely would have directly influenced Chambers to write the bestselling heretical *Vestiges,* which is a book on the same general topic. From that great likelihood, it is possible to see how, at the very least, Matthew probably indirectly influenced Darwin and Wallace, via Chambers, to find further evidence to support his theory, popularize it, and then each lay claim to it as their own independent conception. Those claims of dual independent conceptions being supported by the fallacious premise that no one who could have influenced them, no one such as Chambers for example, had read it before 1858. And we should not forget that Darwin met and corresponded with Chambers before 1858, and that Darwin's mentor, Lyell attended at least one important lecture given by Chambers (Lyell, 2010) Moreover, it is important to remember that before 1858, both Darwin (1847) and Lyell (see Lyell 2010) wrote in private correspondence that they thought they knew that Chambers was the anonymous author of the *Vestiges.* Consequently, a route via Chambers for "Matthewian knowledge contamination" of the brains of both Darwin and Wallace exists. Lyell was in pre-1858 correspondence with Wallace and Darwin, and met with Darwin.

Wallace's (1855) Sarawak paper's editor and publisher, Prideaux John Selby (1842), read and cited *NTA*. Moreover, the eminent naturalist William Jardine had the book in his possession for some time, because he purchased Selby's copy (see Jackson 1992).

The fact that three out of only seven naturalists known currently to date to have definitely read *NTA* pre-1858, played such dynamic roles at the very core of influence and facilitation of Darwin's and Wallace's published work on macroevolution by natural selection might be a mere multiple coincidence. Alternatively, others can argue now that it appears more rational that Matthew's original discovery breakthrough influenced somehow, orally or in writing, both Darwin and Wallace via one or more of these newly discovered published or in-person "knowledge contamination" routes (Sutton 2015).

To labour the point for veracious history, the fact we must not lose sight of in the history of scientific discovery is that the bombshell "New Data" blows to smithereens the old Darwinist excuse, that no one who could have influenced Darwin or Wallace read the original ideas in *NTA* before 1858. Because they did!

Crucially, six of those who cited *NTA* drew specific direct attention to Matthew's discovery of the 'natural process of selection'. They are:

1. Currently unknown — *The Elgin Courier*
2. Anon. — *Edinburgh Literary Review*
3. John Loudon — Publisher, naturalist, botanist, garden designer and polymath
4. Anon. — *United Service Journal*
5. Adam Black — Matthew's Edinburgh Publisher
6. Longman, Rees, Orme, Brown and Green — Matthew's London Publisher.

Seemingly, some six individuals then were so sufficiently cognizant of Matthew's discovery of the hypothetical principle of natural selection that they either went to the trouble of advertising the book on that particular aspect, or else commented on the subject of Matthew's distinctive ideas in this specific area. Moreover, this happened some 27 years before Wallace's and Darwin's 1858 Linnean Society papers, which replicated it, failed to make any known immediate impression on the scientific audience present at their reading. Revealingly, those present at the reading of Darwin and Wallace's (1858) papers apparently failed to notice that anything new, important or distinctive was said on the subject of species. At which juncture, it should not pass unremarked that the only recorded views of those present are of Haughton of Dublin, who remarked (Hindle 1958): *'All that was new was false, and what was true was old.'*

The fact that more of those, who we now know read *NTA*, did not publish any thoughts they might have had on Matthew's theory has two possible explanations: (1) either those who read it did not understand or appreciate the significance of what they were reading, or else (2) they thought it plainly heretical and, therefore, an impossible explanation. Alternatively, once historical accounts of 19th century scientific conventions and norms are considered, the most likely explanation is rather more complex than this simple binary. Before discussing this point further, it is essential to examine additional results of the investigation of Darwin's (1861, 1861a) fallacious 'no one read Matthew's original breakthrough', excuse.

Those Who Definitely Read *NTA* Prior to 1859—Who were also Part of Darwin's Close Social Network *(Black, Jameson, Loudon, Chambers, Murray III, Selby, Johnson)*

Black (1784-1874)

Adam Black was the Edinburgh publisher of *NTA*. Perhaps the first person to read it, he was the son of a master builder. Black started out in business in the same way as Robert Chambers, by selling books. He soon made his fortune and established a significant publishing house, purchasing the publishing rights to the *Encyclopaedia Britannica* in 1829.

In the same year that Darwin finished his first unpublished essay on natural selection, Black ensured that *NTA* was advertised across three quarters of an opening page in the *Encyclopaedia Britannica* (1842), with considerable mention made of Matthew's unique ideas on the topic of species and variety.

Elected lord provost of Edinburgh, and serving two terms of office, Black had earlier served as commissioner of police in Edinburgh. He was a leading member of the liberal Whig Party in 1848, when Robert Chambers stood for election as lord provost. Chambers hoped to be in position to award professorships in the sciences to those whose ideas he favoured (see Secord 2000). Black supported him in that endeavour. However, Chambers met with smear campaigns that more than alluded to the possibility of his authorship of the heretical *Vestiges*. Although Black advised Chambers to fight it out (Secord 2000, p. 295), he capitulated and withdrew his candidacy for fear of the impact the *Vestiges* would have on his publishing empire.

Although we know from their private correspondence that some "in the know," such as Darwin and Lyell, were virtually certain Chambers was the author, (See Darwin 1847 and Lyell 2010) it was not public knowledge. Chambers never responded to allegations that he was. Consequently, public speculation about other possible authors kept the delicious big *Vestiges* secret going in the public eye.

Most significantly, Adam Black, like Chambers, was part of Darwin's social network, as evidenced by what happened when, in 1845, Darwin's best friend Joseph Hooker applied for the chair of botany at the University of Edinburgh.

The professorial appointment Hooker wanted included responsibility for the Royal Botanic Gardens of Scotland, which meant that local politicians had considerable influence regarding who to appoint. Hooker collected some 153 testimonials to support his application, which the Town Council sought to block, since it was not consulted on the fact that the Crown invited Hooker to apply (see Huxley 1918, pp.204-205). Professor Forbes of Edinburgh University forwarded Darwin's letter of support for Hooker to Adam Black, who was then lord provost of the city (Darwin 1845). When Hooker's application was unsuccessful, Darwin was both shocked and quite angry (see Darwin 1845b).

Black's publishing house, A & C Black, moved from Edinburgh to London, where it remains today as part of the Bloomsbury Group, which is the same publisher of Stott's (2012) ludicrous falsehood that after 1860 Matthew never sought foremost priority for his discovery.

There is a bronze statue of Black in Edinburgh's East Princes Street Gardens. I went to see it on the evening of April 10, 2014, following my presentation on Matthew's priority at the International Edinburgh Festival of Science (see Caven 2014). Black was 75 years old when the *Origin* was published.

Jameson (1815–1882)

William Jameson was a botanist, deputy surgeon-general and superintendent of the East India Company. He cited *NTA* in 1853 (Jameson 1853). He cited it to note Matthew's important observation for economic botany that trees could sometimes grow better outside their natural habitat. This was the same original and heretical Matthewian observation that so flummoxed Selby (1842) to cite Matthew and write that he could not understand how it could be so. Selby, a highly religious naturalist (Jackson 1992), would most likely have adhered to natural theology, which saw all the laws of science as evidence of his supernatural God's intelligent design in the scheme of everything, with everything created and positioned precisely in the best location for it and everything else, especially for the comfort of privileged gentlemen humans such as Selby. As Whewell (1833 p.29) wrote in the bestselling Bridgewater Treatises:

'It cannot be accepted as an explanation of this fact in the economy of plants, that it is necessary to their existence; that no plants could possibly have subsisted, and come down to us, except those which were thus suited to their place on Earth.'

Jameson was the garden superintendent at Saharanpur from 1844 to 1875 (Harvard University 2013). According to some accounts, he was not particularly well liked, not a lateral thinker and made some quite notable mistakes during his career with the East India Company (see Rose 2009).

In 1854, the year after Jameson cited Matthew's important scientific observations in *NTA*, William Hooker, Alfred Wallace's mentor and father of Darwin's best friend Joseph Hooker, who was empowered to make such decisions for the East India Company from Kew, blocked his application for promotion in favour of his own protégée, Thompson (see Arnold 2006, pp. 161-162). Jameson's thwarted "due-promotion" was to be superintendent of the Calcutta Garden, the only horticultural gem that was superior to the Botanic Garden at Saharanpur.

Coincidences happen, which is why we have a word for it, and there is zero evidence that Hooker's promotion blocking activity had anything to do with Jameson's citing of Matthew's original and heretical natural selection ideas within *NTA*. However, Jameson's citing of them might have had something to do with the multiple coincidences that in Calcutta, at this very point in time we can find Edward Blyth working. And Blyth was Darwin's most prolific natural history informant (Darwin 1861), whose two most important articles on evolution (Blyth 1835, 1836) were edited by Loudon. In addition, we know Loudon is the same naturalist who had earlier written in 1832 that Matthew had something original to say on 'the origin of species'. The same Loudon that Matthew (1860) told Darwin about in the Gardeners' Chronicle. The same Loudon with whose work Darwin and Joseph Hooker were most familiar. The same Loudon who Darwin cleverly ignored as being a naturalist when he lied with Hooker's approval, that no naturalists (Darwin 1860a) and no one at all (Darwin 1861a, 1861), read the original ideas in *NTA* before 1860.

Blyth was curator of the museum of the Asiatic Society of Bengal. Blyth co-authored two books with Robert Mudie. That fact is relevant because Mudie was apparently first to be second to replicate Matthew's unique phrase "rectangular branching" (see Sutton 2014a for the full details). What all of these possible mere multiple coincidences do reveal is the remarkable extent and degree of integration of naturalists who had direct links to Darwin. For example, Superintendent Jameson's uncle was the celebrated professor of geology, Robert Jameson, who also taught Charles Darwin at Edinburgh University. Most notably, Robert Jameson is widely believed to be the anonymous author of the Edinburgh *New Philosophical Journal* paper of 1826 that praised Lamarck and contained the first English usage of the word "evolved" in relation to the problem of the origin of species.

On first publication of the *Origin,* William Jameson, who remained a regular correspondent of William Hooker, father of Darwin's best friend Joseph Hooker, was 44 years old.

Loudon (1783–1843)

In the scant literature on Matthew, we can see that it was known that the famous naturalist John Claudius Loudon, a celebrated naturalist botanist, garden designer, architect, author, editor and publisher, positively reviewed *NTA* in 1832, deliberately drawing attention to the fact that Matthew's treatise offered matchless insights into the question of the origin of species and varieties (see Clark 1984; Dempster 1983, 1996). Loudon used the phrase "the origin of species," in his review of *NTA*, a phrase that Darwin later used as the title for his famous book. Most notably, the preceding year in 1831, in the *Quarterly Journal of Agriculture,* James Wilson used the same phrase in an article that was critical of recent speculations on the origin of species via one species acting upon another. Given the timing, same notion of one acting upon another to create new species, and the fact that the article was in an agricultural journal, it seems quite likely that Wilson could have been referring to *NTA*.

Just how significant this finding is, I am not sure. For some currently un-established reason, "origin of species" is the only phrase among several hundred that I have looked for that simply does not work very well using IDD. Consequently, it is impossible to discover much about its early usage. Given this weird, apparent Big Data glitch, the earliest use of the phrase I was able to find is in a 1789 book by George Law. However, according to Stott (2012, p. 122), the phrase "origin of species" dates back to at least 1722, where it was deployed in an early draft of Maillet's *Telliamed*, which was doing the rounds of publishers in Paris (see de Maillet 1968, for the most faithful translation). The Term was used also by Alexander von Humboldt (1822), whose work inspired both Darwin and Wallace (Eiseley 1979). Importantly, Humboldt makes it abundantly clear that he was well aware of the dangers of publishing heretical explanations for organic evolution in the first half of the 19th century:

'Whatever relates to the origin of species, to the hypothesis of a variety become constant, or a form which perpetuates itself, belongs to problems in zoonomy, on which it is wise to avoid pronouncing decisively.'

Known as "The Father of the English Garden," Loudon was the son of a farmer. He ran an experimental farm from which he made a considerable profit. Later, he designed St Peter's Square in Hammersmith, London.

Loudon designed the Derby Arboretum, which, some say, served as the inspiration for New York Central Park. Most notably, however, through his great professional influence on Joseph Hooker of Kew Garden's fame, Loudon's Derby Arboretum served as a model for the Royal Botanical Gardens at Kew.

A prolific author and fellow of the Linnean Society, and a corresponding member of the Royal Swedish Academy of Sciences, Loudon was a friend and correspondent of William Hooker, and co-published with William Hooker's close friend and fellow economic botanist John Lindley. Both Hooker and Lindley had their own works reviewed in the exact same volume in which Loudon (1832) reviewed Matthew's *NTA*.

Through the Gardeners' Chronicle, of which he was chief editor at that time, Lindley received a letter from Matthew proving that he was first to import and propagate giant redwood trees in Britain. Lindley would soon claim that Lobb did so. By way of that fallacious assertion, Lindley stole Matthew's glory in the botanical literature (see Sutton 2015). We can see this by that fact that in 1860, Lobb and Lindley were celebrated in 'Knights Pictorial Gallery of Arts' and many other prominent publications, following the mock-up of a huge giant redwood in the world famous Crystal Palace Exhibition, which used the entire bark skinned from a specimen tree. The following year, Darwin (1861a) would successfully portray Matthew as nothing more than an obscure writer on forest trees.

Since we know they read them before 1858, the newly discovered fact that Loudon both edited and published Blyth's (1835, 1836) influential articles on evolution is just one more route for potential probable Matthewian "knowledge contamination" of Darwin's and Wallace's brains.

Aged 60, financially overextended through self-publishing an incredibly expensive natural history book, Loudon died in poverty 16 years before publication of the *Origin*.

Chambers (1802–1871)

We know that the naturalist, publisher, encyclopaedist, geologist and, most notably, the author of *The Vestiges of Creation*, Robert Chambers, read *NTA* because work from it on the training and pruning of trees was mentioned and it was then cited in his typical style on March 24, 1832, in the journal he co-edited and had just established with his brother only six weeks earlier. Both brothers shared the writing workload, but Robert dealt with the leading articles. William's role was to manage the business and compose just some of the scientific papers (Millhauser 1959, Secord 2000).

Chambers anonymously published the *Vestiges of Creation in* 1844. That is the same year Darwin penned his second unpublished private essay on natural selection. The world's leading expert on it (Secord 2000, p.460) describes the *Vestiges* as "the most widely discussed work on science ever published". Darwin's friend Huxley savagely reviewed the book for allowing the hand of God to play a role in evolution. All the while, Darwin, read it avidly and made copious notes (Eiseley 1959). Later, it was one-sidedly unremarked upon by Huxley that Darwin (1861) included the same role of the hand of 'the Creator', in his *Origin*.

Chambers and Darwin met and corresponded. We know that Darwin was aware, as early as early as 1847, that Chambers was the secret author of the heretical *Vestiges*. Chambers even gave Darwin a copy, leading him to write to Joseph Hooker (Darwin 1847) that he believed he knew Chambers was its secret author. And we know that Matthew's Scottish publisher, Adam Black, another who we know surely must have read *NTA,* mentored Chambers in his political career.

According to Millhauser (1959, p.29), by the 1840s, Chambers's home became the meeting place for most of the literary notables of Edinburgh and its scientific community. He forged a particularly close association with the naturalist Edward Forbes, the physicist Dr. Neil Arnott, the chemist Dr. Samuel Brown and the anatomist and physiologist Sir Charles Bell. The possibility that he would not have discussed *NTA* with any one of them seems somewhat unlikely. After all, the very purpose of such meetings was to meet in private to discuss such novel ideas.

In 1840, for his work on geology under the mentorship of Bell (see Millhauser 1959), Chambers was elected a member of The Royal Society of Edinburgh. In 1844, he became an elected fellow of the Geological Society of London, which was the year he published the *Vestiges.*

Millhauser did not know that Chambers read and cited *NTA.* However, he thought it highly likely that Chambers actually knew Matthew (Millhauser 1959, p.82):

'As for Patrick Matthew, his Naval Timber had involved him in a feud (over methods of transplanting) with Chambers' friend Steuart of Alanton, whose own work on arboriculture the Journal had reviewed; it is thus altogether probable that he knew Matthew too.'

Whether he met him or not, Chambers was most certainly interested in following Matthew's work. Because in 1840, we find him citing Matthew's second book, *Emigration Fields,* regarding the ill-effects of tobacco smoking.

Chambers was 57 years old when the *Origin* was first published. And soon after, it was he who famously convinced Huxley to support Darwin's *Origin* in the legendary June 29, 1860 debate against Wilberforce (see Zimmer 2003, pp.62-63).

When Darwin asked him to review the *Origin,* Chambers (1959) replicated Matthew's unique phrase "natural process of selection." He was apparently the first to do so (Sutton 2014a). That was a most mysterious action, which may have been undertaken for a variety of currently unknown reasons.

When Chambers died in 1871, Darwin wrote to his daughter Annie Dowie to express remorse that he had behaved badly towards his scientific work (see Priestley 1908).

Johnson (1799–1878)

Cuthbert William Johnson, like his Bromley neighbour Charles Darwin, was a fellow of the Royal Society. Darwin was also a personal correspondent of Johnson's younger brother, the famous gardener George William Johnson.

Cuthbert Johnson is well known as an agricultural chemist, barrister at law, agricultural writer, public health reformer and sanitary reformer. He was apparently first to be second in print with Matthew's exclusive natural selection phrase "adapted to prosper," without citation, in the *Journal of the Royal Agricultural Society* (Johnson 1841). The following year, in *The Farmer's Magazine*, he cited *NTA* on a different topic (Johnson 1842). The first e-book edition of *Nullius* (Sutton 2014a), contains the full and detailed analysis of all those who were apparently first to be second with apparently unique Matthewisms. Incidentally, to digress at this juncture on this topic, I have an email in my possession, sent to me by a journalist. That email reveals how my work on apparently unique Matthewisms, and the related 'first to be second' (F2b2) hypotheses, led one highly renowned Darwinian historian to recommend a book review article (Malec 2015) of the first e-edition of *Nullius* to Scottish journalist Michael Alexander of the Courier newspaper. The very same historian, who so keenly recommended Malec's article to the press, is also on public record in the press a couple of days later for completely rubbishing the very journal that had only just published it! From these clues, you might work out who the contrarian historian is and what his motives are, but I will not name him here for this. I may, however, name him for it in another book on bias, in the future.

Malec's (2015) review claims that there 'is no Darwin's greatest secret.' But his claim rests solely on his own failure, despite his best efforts, to disconfirm all but one of the 30 examples provided of those apparently first to be second in print with apparently unique *NTA* Matthewisms. I must genuinely thank Malec, nonetheless, for originally disconfirming my claim that Wilkin was apparently first to be second with the term 'figure is best accommodated'. Thankfully, the journal publishing Malec's review gave me a right to reply (see Sutton 2016). By way of that reply, I explain that the F2b2 hypothesis is only a minor part of my research into Matthew, Darwin and Wallace. I make it clear that the discovery of major naturalists, known to Darwin and Wallace, who did read Matthew's breakthrough pre-1858 is Darwin's greatest secret. I then go on to remark how interesting it is, nevertheless, that Malec could disconfirm only one out of the 30 F2b2 cases presented. I certainly expected there to be far more disconfirmations of it than that by now. Anyway, getting back to less recent history about our apparently F2b2 naturalist Cuthbert Johnson, who actually, tellingly, later cited *NTA* pre-1858.

Johnson was a member and prizewinning essayist of the Royal Agricultural Society. He published an incredible array of books on agricultural matters, including a farmer's encyclopaedia in 1844, a wonderful children's spelling book using agricultural phrases in 1846, a book on labourers cottages in 1847, a published lecture on sanitary improvements in 1852, and another on public health in 1852. He also published several important books on fertilizers.

Johnson was at the centre of agricultural publishing and research. In 1832, he was a joint founder of the *Mark Lane Express and Agricultural Journal*, which was politically Whiggish, and, like Matthew's writing, directed toward the tenant farmer. On first publication of the *Origin*, Johnson was 60 years of age.

Murray III (1808-1892)

In 1836, John Murray III formally joined the family publishing house as co-partner with his father, John Murray II. He took over as sole owner following his father's death in 1843. Although Murray III published Darwin's *Origin*, as well as other works by Lyell, he secretly published criticisms of their theories under the pseudonym "Verifier" in *Scepticism in Geology* (1877). Murray's *Quarterly Review* (1833) published a book review of *NTA*. Who actually wrote it may never be known for sure, although diligent archive research in the John Murray archive might well enable intrepid researchers to find out. It is not completely beyond the bounds of possibility that Darwin's friend Lyell wrote it, because John Murray published Lyell's famous *Principles of Geography* in the early 1830s. Lyell was an active book reviewer for the *Quarterly Review* (Carpenter 2008). Clearly, more research is needed in the un-scanned physical document archives of all those who feature in this story.

A letter from Darwin to Lyell (Darwin 1859a) reveals that Murray was not at all happy with Darwin's use of the term "natural selection," and wanted to know from where he got it. In the absence of disconfirming evidence, it seems likely that Murray II (1778-1843) was at least aware of Matthew's use of the longer term "natural process of selection," for the same theory. It is possible that John Murray publishers retained the review copy of *NTA* in the library at their offices. As mentioned above, it is also possible that in 1833,

Murray III may have in some way been directly involved in the 1833 *Quarterly Review* article, since at the time he was active in the publishing house. However, Murray III, who was 51 years old when the *Origin* was published, thought natural selection a ludicrous theory, akin to the notion of a poker mating with a rabbit (Carpenter 2008).

Selby (1788–1867)

Prideaux John Selby was a wealthy landowner, mine owner, quarry owner, agriculturist with a 642-acre country estate and fine manor house. He was a journal editor, a magistrate and High Sheriff of Northumberland, naturalist, farmer, botanist, ornithologist, entomologist, natural history artist and illustrator of works on British ornithology and forestry.

In his own book on British forest trees, several times Selby adapted Matthew's apparently unique phrase "greater power of occupancy" to "great power of occupancy" and was apparently first to be second once with Matthew's full original version (Selby 1842, p.391), where he cited Matthew (1831) and revealed his apparent lack of understanding of one of Matthew's key concepts of natural selection. In that same publication, Selby positively cited *NTA* no less than 23 times.

For what it is worth, in his notebook of books to read and books read, Darwin recorded that he read at least two of Selby's other books. There is, however, no record in what survives of the pages, or non-obliterated text, of Darwin's frequently torn apart notebooks that he read Selby's magnum opus on forest trees.

Well known as an ornithologist, Selby was equally passionate about the forest trees on his estate. As an arborist, he grew them as ornamental specimens in his landscaped gardens and their value elsewhere, for commercial purposes. After receiving a copy of Loudon's *Arboretum et Fruticum*, which cites *NTA*, from his famous ornithologist friend Sir William Jardine, he wrote, on July 24, 1840, asking Jardine to get him a copy of *NTA*. Jardine did so, and Selby retained the book in his personal library (Jackson 1992). Selby appears to have anticipated difficulties obtaining the book in Northumberland (see Jackson 1992, p.86), which is confirmatory evidence the book was not readily available in that region of Britain. He wrote:

'Look out for me a copy of Matthews [sic] treatise on Naval Timber, and a copy of T. Lauder's edition of Gilpins Tree Scenery, as I want both for reference

just now. I take it they were both published in Edinburgh and therefore I think you may be able readily to meet with them.'

I have no idea whether or not Jardine read *NTA* before sending, or perhaps personally handing, the requested copy to Selby in 1840, which was, incidentally, two years before Darwin completed his first unpublished essay on natural selection. It would perhaps be going too far to say for sure that he must have, but I suspect he did. In Jardine's place, I would have read *NTA*, because I am interested in natural history. And I feel sure many, who are themselves interested, would assume that an expert such as Jardine surely did read the book. Besides, there were considerably fewer books in circulation in the first half of the 19th century than later, and because each was printed and bound by hand, they were expensive, luxury items. In sum, it's hard to believe that a curious naturalist such as Jardine would not have taken the opportunity to read the intriguingly important book on trees that Selby asked him to obtain. However, in anticipation of fair criticism of over-speculation, I have, of course, not included Jardine as a known *NTA* reader. We do know, however, from the online Darwin Correspondence Project, that William Jardine was also one of Darwin's correspondents.

In his letter to Darwin of December 20, 1859, we learn that Darwin sent Jardine a review copy of the *Origin*. Sadly, like so much of his correspondence, Darwin's prior letter to Jardine is lost, and there appears to be no other surviving correspondence between them. Notably, however, Darwin's notebook of books read and books to read is absolutely jam-packed with references to Jardine's prestigious publications. Jardine was also a co-editor with William Hooker of *The Magazine of Zoology and Botany*.

The least we can say about the famous naturalist William Jardine is that he held, and for some time kept in his possession, the book that Darwin claimed no naturalist had read. And, just like the famous naturalist Selby, and the naturalist Jameson, and the naturalist Loudon, and the naturalist Chambers, he was closely networked with Darwin's closest social circle. We must thank Big Data technology for allowing me to make these important new discoveries.

A related, but purely speculative, point here is the mere possibility (not probability) that Hugh Strickland, Darwin's mentor and correspondent, was made aware of *NTA* by Jardine and then read it; given that Jardine knew Strickland. Indeed, Jardine's daughter, an excellent ornithological artist, married him!

Strickland died in 1853, six years before the publication of the *Origin*, when he accidentally stepped into the path of one train in order to avoid another. He is renowned for possessing, via Darwin, who in turn got it from Fuller, one of the eight famous Galapagos Islands Fuller finches that were used in several Darwinist reconstructions of events in order to fuel the pervasive myth that Darwin, while on the *Beagle*, discovered the principle of natural selection from observing the local adaptation of the beaks of island finches to best suit local food sources (see Pearn 2009). Darwin did not make the original discovery. Let me explain:

The voyages of the *Beagle* ended in 1836. After his return to England in 1836, Darwin never left the UK again. Two editions of the *Voyages* were published by Darwin (Darwin 1839, 1845c). Darwin carefully edited the second edition of his *Voyage of the Beagle* book (Darwin 1845) by inserting some text on his later observations on finch beaks and evolution. That act made it look as though the thought occurred to him on the voyage (Sulloway 1984). Martinez (2011, p.96) explains:

'The popular myth that the Galapagos finches crucially inspired Darwin to think about evolution arose because in the second edition of his Voyages of the Beagle he added one sentence about finches: Seeing this gradation and diversity, in one small intimately related group of birds, one might really fancy that from an original paucity of birds in this archipelago, one species had been taken and modified for different ends.' But that brief comment was foreign to Darwin's travel books and thousands of research notes; there is no evidence that it represented his thoughts during his voyage in 1835.'

As Martinez (2011) goes on to explain, by the time Darwin snuck that revision into the second edition of the *Voyages of the Beagle*, he had already believed in evolution for eight years. Martinez (2011) provides an excellent account of Darwin's doctoring of the second edition of the *Voyages of the Beagle*. This editing act is arguably a key ingredient of the success of his later plagiarism by glory theft science fraud.

In actual fact, Darwin did far more than subtly sneak in the odd sentence, or odd comment—an impression that one might get from reading Martinez alone. When we visit the primary sources, we can see that Darwin added huge amounts of new text into the second edition of the *Voyages of the Beagle*, without informing his readers that he had done so. The excellent website of the Rockville Press provides a superb comparison of the text between Darwin's 1839 and 1845 *Voyages* by way of presenting comparative text from the Project Gutenberg digitized versions of the two editions in question.

Darwin (1839):

'A group of finches, of which Mr. Gould considers there are thirteen species; and these he has distributed into four new sub-genera. These birds are the most singular of any in the archipelago. They all agree in many points; namely, in a peculiar structure of their bill, short tails, general form, and in their plumage. The females are gray or brown, but the old cocks jet-black. All the species, excepting two, feed in flocks on the ground, and have very similar habits. It is very remarkable that a nearly perfect gradation of structure in this one group can be traced in the form of the beak, from one exceeding in dimensions that of the largest gross-beak, to another differing but little from that of a warbler.'

Darwin (1845):

'Of Cactornis, the two species may be often seen climbing about the flowers of the great cactus-trees; but all the other species of this group of finches, mingled together in flocks, feed on the dry and sterile ground of the lower districts. The males of all, or certainly of the greater number, are jet black; and the females (with perhaps one or two exceptions) are brown. The most curious fact is the perfect gradation in the size of the beaks in the different species of Geospiza, from one as large as that of a hawfinch to that of a chaffinch, and (if Mr. Gould is right in including his sub-group, Certhidea, in the main group) even to that of a warbler. The largest beak in the genus Geospiza is shown in Fig. 1, and the smallest in Fig. 3; but instead of there being only one intermediate species, with a beak of the size shown in Fig. 2, there are no less than six species with insensibly graduated beaks. The beak of the sub-group Certhidea, is shown in Fig. 4. The beak of Cactornis is somewhat like that of a starling, and that of the fourth subgroup, Camarhynchus, is slightly parrot-shaped. Seeing this gradation and diversity of structure in one small, intimately related group of birds, one might really fancy that from an original paucity of birds in this archipelago, one species had been taken and modified for different ends.'

The first (1839) edition of the *Voyages of the Beagle* contained no such clue that Darwin thought about natural selection while on the Beagle expeditions. In reality, on the Beagle expeditions Darwin believed, and continued to believe until around 1837-39, that species were immutable.

Those finches, often called "Darwin's finches," were collected by his shipmate, who was Captain FitzRoy's steward (Harry Fuller). And for years after his return to England, Darwin saw no significance in those finches. He simply believed that they, like all species, were immutable. It was not until he got back to England and started reading books, no less, that he became an organic evolutionist.

Contrary to pervasive Darwinian myth mongering on this subject, it would be over 100 years after Darwin's return from the *Voyages of the Beagle* before scientists worked out the natural selection significance of Galapagos finch beak adaptations. As Sulloway (1982) proved:

'Darwin identified the cactus finch as an 'Icterus,' a genus in the family of orioles and blackbirds, and he mistook the warbler finch for a 'wren' or warbler. In fact, Darwin correctly identified as finches only six of the thirteen species—less than half the present total—and he placed these six species in two separate groups of large-beaked and small-beaked Fringillidae. Furthermore, with the exception of the cactus and warbler finches, Darwin failed to observe any differences in diet among the various species, mistakenly believing that their diets were largely identical

For this reason he could never argue that the different beaks of these finches were necessarily adaptive and therefore produced by natural selection. Thus there is no basis to the claim that Darwin had these finches in mind when he broached an evolutionary interpretation of the mockingbirds and the tortoises in his Ornithological Notes.'

Finches are mentioned just twice in the first edition of the *Origin of Species* (Darwin 1859), but neither of the two references made to finches is on beak adaptation between different types of finch.

Perhaps one reason why finches and all their different beaks feature so largely in Darwinist mythology is because of a book published in 1947 (Lack 1947 see Marx and Bornmann 2013), which created the myth of "Darwin's Finches" to fill in the knowledge gap of Darwin's missing Eureka moment. In this 1947 book the term "Darwin's Finches" was coined. Lack (1947 xiv) wrote in his preface that: *'Charles Darwin appears not to have appreciated the evolutionary evidence provided by the finches until several years after his return from the islands.'*

The truth is worse than that, however. Darwin never wrote anything at all worth reading about those Galapagos finches, due to his dismal failure to as much as note which birds came from which islands. Consequently, most of what he did later write about them was an absolute mess of assorted errors (see Sulloway 1982).

Here then, we have yet another example of the "Darwin Eureka

Moment Myth" being created, like all knowledge gap myths, simply to fill the knowledge gap regarding when exactly he is supposed to have plausibly discovered the law of natural selection.

Most notably, seven of the eight finches originally collected by Fuller were passed to Jardine, no less!

Once again, therefore, and with excruciating irony, we find direct links to *NTA* right at the very centre of Darwin's pre-1858 social universe.

Strickland, the geologist and naturalist who, incidentally, we know owned the last (8th) Fuller Finch, shared Selby's passion for ornithology. Because they were close friends, Strickland was a frequent visitor to Selby's home (Jackson 1992), and so, presumably, as a fellow gentleman of science, was allowed into his library. In which case, whether alerted to *NTA* by Jardine or not, we know that Strickland was probably at least within arms-length of that book for more than an appreciable moment, on more than one occasion.

Strickland was elected a fellow of the Royal Society in 1853, and was a regular correspondent of Darwin. He led the team that included Darwin, which drew up the first formal codification on the rules of scientific priority for the British Association for the Advancement of Science. He engaged in some lengthy correspondence with Darwin (1849), who tried to get the rules changed so that originators would lose priority to more famous naturalists, such as him, who worked out more of the details of their discovery. Obviously, given what we now know, this Darwin and Strickland priority affair is a highly relevant topic in the story of Matthew and Darwin.

Of all those in Darwin's close social network who we know read *NTA,* Selby, a fellow of the Royal Society of Edinburgh, was the most integrated, being closely associated with William Hooker, (see Brock and Meadows 1998), Charles Lyell, Thomas Huxley and, most importantly, with Darwin by way of their mutual senior capacities at the British Association for the Advancement of Science, Royal Society and the Linnean Society.

Selby and Darwin's friend and champion Huxley were also members of the Ray Society, founded by Strickland. Selby's and Darwin's association with William Hooker, and Hooker's association with Wallace, may well have had some influence on Wallace, sending his 1858 Ternate paper to Darwin and the consequent reading of their papers before the Linnean Society that year. What makes this seem like a possibility worthy of further research is that Selby was editor of the journal that published Wallace's Sarawak paper in 1855.

Notably, eight months after Wallace's paper was published in the September 1855 issue of Selby's journal, Lyell purposefully visited Darwin in order to persuade him into publishing his research on the "Natural Selection Theory" sooner rather than later.

Exactly how the connection between Wallace, Selby and Jardine came about, and whether it can be proven that they discussed *NTA*, is an important question in need of further research in various archives of their respected journals and correspondence. Whatever the outcome of such future research, once again the pernicious Darwinist myth that Matthew's book was obscure, unread by any naturalist and had no possible or plausible influence on anyone who could have influenced Darwin or Wallace is not just bust. It is blown to smithereens by the bombshell original discoveries unveiled in this book.

Besides Strickland, other guests at Selby's house, all of whom would have been within arms reach of *NTA* within his library, and were personal correspondents of Darwin, include John Gould, Leonard Jenyns, William Yarrell and, of course, Sir William Jardine (see Jackson 1992), and also Darwin's father, no less, another member of the Royal Society.

A founding member of the British Association for Advancement of Science, Selby attended his first meeting in 1833, and his best friend William Jardine was also a member, chairing the Zoological section at many meetings (Jackson 1992). Selby initially joined William Hooker—a friend of Darwin's, and father of Darwin's best friend and botanical mentor, Joseph Hooker— as a founding editor of *The Magazine of Zoology and Botany*, which later became known as the *Annals and Magazine of Natural History*. And we know that the library William Hooker directed at Kew did at least later hold a copy of *NTA* (Royal Botanic Gardens, Kew 1899). Joseph Hooker worked at Kew, and Darwin visited both of the Hookers there.

Selby was a very close friend of Darwin's great friend Leonard Jenyns. The Darwin Correspondence Project has 40 letters that passed between Jenyns and Darwin. Jenyns (1885) wrote a book about Selby, in which he recorded visiting him at his home along with Darwin's father.

Given Selby's obvious enthusiasm for *NTA*, repeatedly evidenced in his citations of Matthew's natural selection concept of "greater power of occupancy" and his obvious respect for its author's knowledge of arboriculture, it seems highly unlikely that he would not have discussed *NTA,* at the very least with other connected gentlemen of science. In order to do such things, scientists get together and establish societies, associations, clubs, committees and standing conferences. At these gatherings, Selby mixed further with Darwin's closest friends, many of whom, as we know, Jackson (1992) reveals were Selby's house guests.

Given that the British Association for Advancement of Science was founded in the very year that *NTA* was published, that *NTA* was one of only seven books on botany published that year and given Loudon's (1832) review of it, which mentioned its originality on the subject of "the origin of species," it is difficult to suppose that *NTA* would not have been a topic of conversation of those forming a society in 1831, to share ideas, advance knowledge and meet with other inquisitive luminaries. After all, one of the 1831 founding objects of the British Association (Rennison 2009, p.109) was "to promote the intercourse of the cultivators of science with one another..." Small wonder then that it was Loudon, an expert cultivator in every sense, who went on to edit and publish Blyth's (1835, 1836) two most influential papers on organic evolution.

Selby also associated with Charles Lyell, in the capacity of being a founding member and vice president of the British Association, while Lyell was a member of its council. By 1854, Huxley (see Leighton 1851) and Joseph Hooker were also members. Hooker also had served on its council. Among the membership of the British Association were Darwin's other associates, Asa Gray and Baden Powell. That the British Association had a long standing Kew Committee—of which all members of the Council, including Darwin, were members—may not be insignificant in understanding who read *NTA,* and with whom they discussed it.

We know that Darwin was extremely interested in trees and birds, particularly pigeons, and that Selby was a published expert on both. And yet, oddly, while Selby's books on pigeons and parrots are listed in the surviving text of Darwin's notebook of *Books Read and Books to Read*, his 1842 book on British forest trees is not. Neither is it mentioned anywhere else in Darwin's published work, nor in any other unpublished documents that are available online. If Darwin did read Selby's 1842 book, and it's hard to believe that he never did, then he never wrote about it, or else he did write about it in some lost letter, unknown essay or notebook, or on one or more of the many pages that are now missing or torn out from his various known notebooks.

Most tellingly, in 1842, the same year in which Selby published his book on British forest trees, he was vice president of the British Association, while Darwin sat on its council. In that same year, the British Association was supporting Darwin and the celebrated American botanist Asa Gray, among others, to conduct research into the races of men, which, most notably, was another important topic discussed in *NTA*, from the standpoint of natural selection. With their shared, close network of friends, mentors and associates, their corresponding interests in birds and trees, economic botany and professional administrative duties within the British Association, it seems, arguably, much more likely than not that Darwin and Selby would have met and would have discussed *NTA*.

Selby was 71 years old when the *Origin* was published, Jardine was 59 and Strickland had been dead six years. Selby, though liberal in his politics, was actively against the Chartist movement. He unsuccessfully contested Berwick as a Liberal in 1812. Unlike the Scottish Chartist regional representative Laird Matthew, his morality was bounded by high self-interest. During the outbreak of Chartist activity in 1848, Selby wrote to Jardine that he was glad the disgraceful behaviour in Edinburgh and Glasgow had been "put down" (see Jackson 1992, p. 8). That said, he was noted as a man who was able to get along well with those who shared opposing views. I strongly suspect, however, that this devout Christian, naturalist, politically Liberal, land owning, mine owning and staunch anti-Chartist, would have been at turns intrigued and perplexed by the mix of politics, news, knowledge, atheism and radical socio-biological ideas running through *NTA*; ideas that he would not wish to promote or to see gain a wider audience among the lower ranks of Victorian society; certainly not when those ideas were written by Matthew, a Scottish regional representative of the Chartist movement (Desmond 1989a).

Debunking Darwin's and Dawkins' Excuses for Denying Matthew's Greatness

That Selby and other 19th century gentlemen of science did not comment in the literature on Matthew's discovery is not likely to stem from their failure to comprehend it, and to repeat the point already made, it was most definitely not a conspiracy of silence. A more likely explanation that we can glean from the work of the historian James A. Secord is that it was a natural and general result of their conventions and norms, which arose under a particular set of emerging conditions that would lead to professionalization and specialization in science. Nonetheless, they were governed in no small part by a cadre of Oxbridge parson-professors like Baden Powell.

Specifically writing about the treatment of Chambers' *Vestiges*, Secord so superbly explains why heresy was given the silent treatment in the first half of the 19th century, that I am obliged, therefore, to quote some of his excellent scholarship at length (Secord 2000, 420-421):

'Science in these circles was embedded in codes of gentility, which meant that claims to legislate over nature were unlikely to succeed. As long as the gentlemen of science expressed their views in the appropriate manner, which often meant sticking to their experience except when speaking in confidence, they could believe what they wished on religious and political issues. By regulating their talk and expressing certainty only for specific 'facts' in their specific 'departments,' men of science could be both polite and authoritative at the same time – something that was not always easy to do. Beyond that, they were no more than ordinary participants in the conversation that defined polite society. The modest social origins of many men of science meant that silence was the most effective way of exercising authority. They spoke to larger issues where consensus already existed ... Neutrality was necessary if science's claims to absolute truth were not to conflict with the demands of civility. To say more would have been inappropriate and boring.'

For his massive breach of etiquette, those same gentlemen of science saw Matthew as just such a bore; a fact revealed in Dempster's indispensable book on Matthew (Dempster 1996, p. 5):

'For the British Association meeting at Dundee in 1912 Calman, a deputy director of the Natural History Museum, was given the duty of presenting some facts about Patrick Matthew's contribution is largely a translation from the

German essay by May (1911). It is clear from a letter sent, prior to the meeting, to D'Arcy Thompson that Calman had little regard for Matthew who is referred to as 'an old bore.'

In the first half of the 19th century especially, men of science were looking to become respected and professionalized, looking to carve out a niche in society for their discipline. As Yeo (1984, p.9) explains:

'...close relationship between science and general cultural debate, together with the insecure status of the scientific community, made the authority of science a significant issue. Scientists had to establish the domain of natural knowledge as their own, and monitor the boundaries between science and religion.'

The surgeon William Lawrence was forced to withdraw his book (Lawrence 1819) for the scandal of its heresy on evolution. Matthew's book was not only equally as heretical it was full of seditious political ideas. Matthew proposed political reform based upon his scientific observations of nature (see Sutton 2014a), including the suggestion that the aristocracy and gentry marry 'fresh stock' from the lower social classes for both physical and moral reasons (See Matthew 1831, p.365-366). In particular, he saw the law of hereditary title to landed seats as a social problem that prevented social advancement by way of disallowing 'natural' competition among those most fitted to reach the top. He predicted this would lead to social unrest *Matthew (1831, p.365)*:

'The law of entail, necessary to hereditary nobility, is an outrage on this law of nature which she will not pass unavenged — a law which has the most debasing influence upon the energies of a people, and will sooner or later lead to general subversion...'

For that same reason, the norms of the Royal Society stated that its members should discuss nothing about God or politics, and news that was unconnected to the business of philosophy should be avoided at all costs (Gleick 2010).

Clearly, Matthew's book broke these brute censorship norms adopted by the Royal Society. That fact alone makes a complete mockery of Richard Dawkins' (FRS) claim (Dawkins 2010) that Charles Darwin (FRS) deserves even greater recognition on the grounds that Matthew (incidentally not a fellow of the Royal Society) should have "trumpeted his discovery from the rooftops" if he truly knew the importance of his own great breakthrough. As a supposed 'leading expert' on the topic, Dawkins seems also weirdly unaware that Matthew (1860b) explained to Darwin that in the first half of the

19th century naturalists feared pillory punishment were they to teach or otherwise disseminate his bombshell breakthrough, and that his book was banned by Perth public lending library. Dawkins, in carrying out his ill-informed, context-free, misleading, cherry-stepping, critique, appears equally weirdly uninformed that the subject of his apparent "expert" ignorance is supported by evidence in the publication record. For example, the United Service Journal and Naval and Military Magazine (1831a, p.457) markedly disclaimed anyone participating in daring to so much as think about Matthew's heretical ideas on the natural process of selection: '...*we disclaim participation in his ruminations on the law of Nature...*"

For what he dared to share in such a God-fearing environment, shouldn't the Scot Matthew be lauded as a national and international hero of science, rather than ignorantly mocked by the world's leading atheist for disappointingly failing to orchestrate his own martyrdom punishment?

Back in the 19th century, the diminishment by silence of Matthew's claim was observed by members of the British Association for the Advancement of Science, an organization which, incidentally, continues the shameless practice at the time of writing. This provides us with a clue to one possible reason, among a multitude of interconnected possibilities, why not a single one of its members, and we newly know for sure that some of them read *NTA*, stood up and challenged Darwin's fallacious excuses and lies for supposedly not having read it. Eiseley (1959, p.176), for example, explains the attitude of the times, in which failure of gentlemen of science and their elite clubs, in the first half of the 19th century, to react to heretical books like *NTA,* was something highly expected of them:

'... *at a time when the multitudinous adjustments of organisms to their environment were evidence of the direct hand of God in earthly affairs, had been vigorously promoted through a long series of theological naturalists from John Ray and William Derham to William Paley... in particular enhanced the feeling of wonder towards the works of God and increased human faith in Divine Providence.'*

Consequently, by privately depicting and publically treating Matthew as a lesser man than themselves, those gentlemen of science were perhaps effectively relying upon a classic guilt neutralizing technique (Sykes and Matza 1957) for the justification of personal wrong doing through adopting officially sanctioned, in-group reasons. If so, those reasons were attributed to the highest divine authority imaginable. Namely, the things other men merely created for them to simply believe about the wishes and actions of their Abrahamic god.

We might hypothesize that Matthew was effectively being punished by religious dogma for taking no prisoners in *NTA*.

Having lived in Germany and France, and being fluent in both languages, it seems likely Matthew would have read the works of Buffon and Diderot while in France. Both of those naturalists had been subject to censure by the Catholic Church. Diderot was imprisoned and interrogated for his work on evolution (see Stott 2012 for a riveting account). In a footnote, Matthew wrote exactly what he thought about such things (Matthew 1831, p.131):

The dread of change in Catholic countries—the proscription of almost every new work treating of science—the complete submission of the mind to the religious authorities, 'bearded men becoming little children' even to the letter—the consequent general abandonment to sensual enjoyment—the immense number of holidays and the shoals of meddling priests are a great bar to improvement an insurmountable one to manufacturing pre-eminence.'

While Darwin, who at Cambridge had studied to become a parson, struggled with his faith in the absolute secrecy of his immediate family, most other naturalists personally identified with the absolute ideal of the Christian gentleman of science as one type above all others whose word was a matter of honour and given, therefore, with unquestionably honesty. It is most likely for this reason that Matthew, a man not normally wary of writing exactly what he believed, was fully cognizant of the dire implications for Darwin, and perhaps for himself facing a writ of libel, if he were to accuse Darwin of plagiarizing and lying in his *Gardeners' Chronicle* letter of 1860. Were things to escalate from such an accusation, then as it is now, the money would be on the suit, and Darwin's pockets were probably deeper than those of Matthew, who was bankrupt following the failure of his Scots New Zealand Land Company. Furthermore, Darwin had many of the wealthy and influential gentlemen of science from the UK and USA on his side. Who did Matthew have?

We can see the rank-closing behaviour of the scientific community against the outsider Matthew by way of their response to a letter he sent to *The Dublin University Review* three months before his published claim to his discovery in the April 1860 Gardeners' Chronicle. The Dublin University Magazine chose simply to mock Matthew as a deluded crank, and to praise Darwin as the heroic originator of the theory of natural selection. The mockery was in an article written by the geologist, Darwin's Cambridge University friend and correspondent, David Anstead (see Sutton 2014a):

'In the Gardeners' Chronicle for 7th February 1860 is a long communication from Mr Patrick Matthew of Gourdie, NB the author of a treatise 'On Naval Timber and Architecture,' in 1831, in which a claim is made by the author to have been the originator of Mr Darwin's theory of natural selection. In a letter to the editor of this journal Mr Matthew has repeated the claim and considers himself wronged by the remarks in our journal of February (vide p 235). We cannot however perceive, either in the extracts from his work, or in his remarks, any thing more than a repetition of a fact long familiarly known, namely that many species pass into each other by insensible gradations—a fact acknowledged by all.

Naturalists, and to account for which, Lamarque's theory of the modification of specific characters was not the first invented. A statement that individuals and varieties were often involved in a struggle for existence, in which the strongest and the best adapted to the circumstances of the moment would prevail—a knowledge of the existence of sporting varieties in many well known species, and the possibility of certain modifications introduced into species in consequence, do not interfere with Mr Darwin's claim to be regarded as the first who has put forward the principle of natural selection as the method adopted by nature to insure a succession of varieties resulting in species adapted to continue throughout all time and in absolute perfection, the chain of created beings.'

The Dublin University Magazine was biased in fighting for Darwin's corner in 1860. It denied Matthew a publication of the letter he sent them in February 1860. Next, it slapped him down on pages 717 to 718, and went even further, earlier on page 32, to explain what everyone who had read it knew, namely that the first edition of the *Origin* was an abstract and that poor Darwin, whilst ill, had heroically written it in a priority-rush for his own personal glory:

'Whoever does set himself up to judge this book would do well to remember that he has before him nothing but an abstract. Mr Darwin hopes in two or three years to be able to complete it. At present he is compelled to omit whole masses of facts and of references to authorities for the several statements advanced. It is perhaps to be regretted that, by publishing this abstract he should invite criticism founded on imperfect data and induce his adversaries to entrench themselves in

positions, from which an instinctive dislike to the humiliation of a palinode may render it difficult to dislodge them. However his health was far from strong, and he found that he was being anticipated by Mr Wallace's researches made independently.'

The bankrupt Matthew might perhaps have calculated in a wink all the angles, their risks and benefits. The embodiment of the ideal of a Regency gentleman, he kept his cool and wrote for Darwin's benefit, a chivalrous face-saving sentence in the *Gardeners' Chronicle*, which made it clear, in print at least, that while he believed Darwin's excuse that he had no prior-knowledge of *NTA*, the "expert" Darwin had, nonetheless, simply got his basic facts wrong by claiming no other naturalist had knowledge of his prior published breakthrough. Here it is again (Matthew 1860b):

The Origin of Species, I notice your Number of April 21 Mr. Darwin's letter honourably acknowledging my prior claim relative to the origin of species. I have not the least doubt that, in publishing his late work, he believed he was the first discoverer of this law of nature. He is however wrong in thinking that no naturalist was aware of the prior discovery...'

Given what the newly discovered data in this chapter reveals about the number of newly discovered naturalists who did read *NTA*, including Loudon, whose work was very well known to Darwin and Hooker, why was Darwin not challenged by others besides Matthew on the falsehood that none had done so? The answer is perhaps a complex one. Maybe it lies in no small part in understanding the ironic unintended consequences of the influence of organized religion, within which Darwin opportunistically machinated.

Secord (2000) explains that in 19th century society, the ideal was that wealthy gentlemen of science did not need to make important discoveries in order to earn a living. Independently inherited income was seen, therefore, as a highly desirable safeguard of objectivity. And the ideal scientist was a man of honour, whose word could be absolutely trusted above that of others. According to modern mythology, Darwin was just such a man. Indeed, according to our mythology he is the epitome of such a man; of such an honest scientist.

Matthew committed a great scientific impoliteness in *NTA*, one that no naturalist could have committed without being duly black-balled from joining the necessary cliques, organizations, clubs and committees, as well as from obtaining prestigious appointments. Namely, he had committed the tri-faux pas of publishing a big deduced idea unsupported by enough examples of confirmatory facts, which then unforgivably trespassed upon and directly challenged the most important pastures of Christian revelations of divine truth about the origin of species. Finally, to add insult and fear to injury, he mixed into it his radical libertarian political beliefs.

Ironically, in the *Origin,* unlike Matthew, Darwin, the man that Dawkins and so many others unquestioningly worship as their hero of atheism, shrewdly cowered away from writing about the issue of humans and natural selection. This is evidenced in Darwin's own words from 1859, as paraphrased by Eiseley (1959, pp.256-256):

'Nothing better illustrates the oppressive theological atmosphere of the time than Darwin's response to an inquiry from Wallace prior to publication of the Origin as to whether he intended to discuss man. Darwin rejoined as follows: 'I think I shall avoid the whole subject, as so surrounded with prejudices, though I fully admit it is the highest and most interesting problem for the naturalist.' In a similar vein he confessed to Jenyns: 'With respect to man, I am very far from wishing to obtrude my belief; but I thought it dishonest to quite conceal my opinion.'

In the clamour that arose after the *Origin* was published, Darwin could not avoid insinuations of deceit by way of observations made on his failing to elaborate upon the place of man in "his" system. It was perhaps partly in indirect answer to such slurs that Darwin undertook authorship of *The Descent of Man* in 1871. He did so, however, when his position as originator of macroevolution by natural selection ceased to appear so novel and revolting to the public mind.

In the judgment of the present writer, there can be no doubt, considering the temper of the times, that Darwin's caution was well justified and probably had the salutary effect of broaching what was then an unpleasant topic by successive doses, which were found more acceptable by increment rather than, as Lyell was accustomed to saying, "going the whole orang" at once.

Remarkably, 11 years after the first publication of *NTA*, and in the same year that Darwin completed his first unpublished private essay on natural selection, Selby went into print with no more than the following about Matthew's fully heretical theory, and cited him accordingly (Selby, 1842, p.391):

'The soil upon which most of the Abietina prevail, is usually of a dry and cool quality; thus the debris of granitic and other primitive rocks and barren sandy districts are very commonly occupied by Pine and fir forests, sometimes of enormous extent; the thick and close manner in which they grow, and the dense shade they produce, effectually preventing the vegetation of other species. Matthew, however, in his able treatise on naval timber seems to think that its indigenous location in such districts arises not so much from preference of soils of the nature above mentioned as from its having more power of occupancy in such soils than any other plant of the country; and this opinion he endeavours to support by stating that the Pinus sylvestris, planted in a good or rich soil, attains larger dimensions and its best timber properties, and that it is only driven from this superior soil by the greater power of occupancy possessed by the oak and other deciduous trees, an opinion in which we cannot altogether acquiesce, as we see no reason why the fir, if it grows with such additional vigour in a richer soil, as Mr Matthew asserts, should at the same time be unable to maintain a contest with the oak or other trees.'

Here then we see an example of exactly what Secord (2000) is talking about in terms of a gentleman scientist of the 1840s keeping himself on safe ground by politely criticizing Matthew on a point of botanical and arboricultural expertise. By so doing, Selby has managed to criticize one important element of Matthew's natural process of selection without mentioning it outside of the practical issue of Selby's apparent failure to understand why there might be a multi-factored natural process of selection in nature. By means of such selective silence, Selby is able to criticize Matthew's treatise without engaging with its heretical dissent on the question of intelligent design determining the origin of species and where they are located on the planet for the benefit of humans.

I cannot help wondering about the possibility that *NTA*'s happy to accept criticism tone made some meaningful connection with Selby, a fellow arboriculturalist, on a personal level, which might have prompted him to break with the complete silent treatment and engage with that one relatively safe aspect of Matthew's theory. In short, was Selby prompted to write by the last words in *NTA* before the Appendix (Matthew 1831, p.359)?:

'As a friend, we have stood on no ceremony with our brother arboriculturists. We have laid ourselves open to their criticism, and we hope they will shew as little ceremony with us.'

How much richer would scientific progress have been in the 19th century if naturalists had been culturally allowed to discuss *NTA* in print as Matthew wished? We cannot know. But what Selby, for one, missed so engaging with was Matthew's clearly articulated pronouncement of the fact that simple binary explanations, of the kind Selby sought, though more likely to be popular because they are easy to understand, remember and disseminate, are not necessarily veracious. For example, in the crowded natural forest, one species towers over others, some species do well in the lower, shadier regions, while others rot and wither. For decades, or centuries, in Britain, species such as pine or beech might predominate, only to lose dominion following a period of drought or disease, thus allowing competitors such as oak to take over (Green and Ray 2009). In a country with a mild climate, such as Britain, with droughts every 30 or 40 years, oak trees probably do enjoy a power of occupancy in soil well suited to fir trees. In areas with much hotter summer climates, such as the south of Italy, for example, naturally selected species of pine thrive and predominate where no currently existing variety of oak can possibly compete.

So close to this striking and multi-level complex aspect of nature, none could be better circumstance suited than Matthew, the lateral thinking politician, businessman, farmer, botanist and forester, to see and comprehend de Candolle's (see Lyell 1832, p.131) understanding of the role of ecological footholds in species competition and its explanatory significance regarding the survival of the best circumstance suited trees as a way to understand outcomes of wider competitive struggles in nature and human society.

Matthew, unlike so many others in his time, fully understood that in nature, as in society, sometimes the downtrodden are just in waiting for the right circumstances to happen. After which, they overtop their rivals and take over. His analogy between society and nature exposed how, under the artificial selection of Western human culture, an artificially maintained greater power of occupancy of the landed gentry and other inherited wealth and rank of the privileged classes, was preventing members of the lower social orders from attaining their full natural potential in environmental, technological and intellectual circumstance, where they would otherwise thrive.

In sum, this discovery in the literature of Selby's incredibly limited matter of fact criticism of Matthew on the sole issue of the species competition between oak and fir trees does not disconfirm Secord's

(2000) general explanation regarding 19th century permissible topics of published critical enquiry.

Matthew's description of the constant, competitive struggles between different species, leading to the establishment of impermanent ecological niches, should have been extremely enlightening for Selby and his contemporaries. Were it not for the fact of its religious heresy and dangerous justification for the future Chartist political unrest that followed, published in a book on naval timber—the growing and obtaining of which was of the highest national importance and patriotic duty for shipbuilding, housing and industry (see Nail 2008)—at a time of great social unrest, when there existed powerful social strictures against gentlemen of science commenting upon the subject of the origin of species, I suspect Selby and others would have trumpeted Matthew's discovery from the rooftops.

As it was in the first half of the 19th century, the British upper classes feared their own underclass might rise up against them, as happened in France. That made Matthew a dangerous provocateur.

Eiseley believed that Adam Sedgwick's Presidential Address before the Geological Society of London was a criticism of Lyell's heretical proof of the greater than biblical age of the Earth, but it might, for all we know, have been directed at Matthew as well (Eiseley 1959, p.266):

The more religious-minded and the more sober-headed continued to cling to the views expressed by Adam Sedgwick in his Presidential Address before the Geological Society of London in 1831, just about the time young Charles Darwin was departing upon his memorable voyage.'

Sedgwick's speech, made in the very same year *NTA* was published, may have included an attack on the uniformitarian hypothesis of Sir Charles Lyell (1830) that from the deep-time pre-history to the present, all geological changes happened constantly and at a uniform rate, but also it criticized any attempt to explain the appearance of man by laws of nature. Lyell at that time believed in divine creation of all species, while Matthew's book went further, relying on Lyell's excellent dissemination of evidences, gathered by others for the greater age of the Earth, to hypothesize a new law of nature. Matthew's theory could explain the varieties of man. In relation to the origin of new species including humans, he explained it all through a process of naturally selected variation. Matthew even went so far in the Appendix to *NTA* to vote for natural selection over divine creation as the best explanation for the origin of species.

Matthew accepted Lyell's provision of evidences for the Earth being many millions of years old, and he accepted Lyell's notion that there were great periods of geological stability that were characterized by gradual change. However, Matthew understood the truth of something Lyell rejected. Matthew saw that major, yet rare, geological catastrophic events had occurred throughout the Earth's history. Matthew saw how these events had a major impact on the emergence of new species by natural selection to be better circumstance suited to the change in circumstance. He also understood that natural selection continued to occur as a process during great periods of geological stability.

When Lyell followed Matthew in print a year later, he made reference to a conclusion formed regarding an entirely hypothetical source of evidence of new botanical species formed in a garden (Lyell 1832, p.56):

'...*we have no data as yet to warrant the conclusion that a single permanent hybrid race has ever been formed even in gardens by the intermarriage of two allied species brought from distant habitations. Until some fact of this kind is fairly established, and a new species capable of perpetuating itself in a state of perfect independence of man, can be pointed out, we think it reasonable to call in question entirely this hypothetical source of new species. That varieties do sometimes spring up from cross breeds, in a natural way, can hardly be doubted, but they probably die out even more rapidly than races propagated by grafts or layers.*'

Who else on Earth other than Matthew could Lyell have been referring to in 1832? If anyone besides Matthew (1831) was the original source of such a brand new theory for limitless organic alteration in nature, explained by way of analogy to artificial selection by man, I don't think we have yet discovered them in the literature. Given Lyell's reference to hybridization in a garden, I cannot help wondering whether just maybe Laird Lyell was thinking about his Scottish neighbour, Laird Matthew's, shabby treatment on the front page of the Edinburgh Literary Journal (1831):

'*Mr Patrick Matthew as we understand is a small landowner on Gourdie hill, near Errol, in Perthshire, an inconsiderable orchardist, if we may so speak, who has a house, with a garden, and shrubbery, where he makes experiments on fruit trees.*'

If Matthew, like Darwin after him, depended upon Lyell's (1830) first volume of his three-part set of the 'Principles of Geology', it seems that the great Lyell may just possibly have depended upon Matthew's (1831) book of the following year to inform his second volume of the year after that, because he not only replicated Matthew's concept of greater power of occupancy, he also used the same word, "encroachments," that Matthew used, and in the same context (Lyell 1832, p.156):

'Every species which has spread itself from a small point over a wide area, must, in like manner, have marked its progress by the diminution, or entire extirpation, of some other, and must maintain its ground by a successful struggle against the encroachments of other plants and animals.'

Only the year before, Matthew (1831, p.387) had written much the same thing, using humans as an example:

'As far back as history reaches, man has already had considerable influence, and made encroachments upon his fellow denizens, probably occasioning the destruction of many species, and the production and continuation of a number of varieties or even species, which he found more suited to supply his wants...'

Most tellingly, Darwin's 1842 private essay contains the exact same "supply his wants" expression regarding artificial selection:

'By such selection make race-horse, dray-horse—one cow good for tallow, another for eating &c.—one plant's good lay... leaves another in fruit &c. &c.: the same plant to supply his wants at different times of year.'

On page 116 of the *Origin* (Darwin 1859), we should note that Matthew's notion of the importance of the "power of occupancy" to defend or succumb to encroachments as a way to explain both extinction and evolution is entirely replicated by Darwin:

'...the modified descendants of any one species will succeed by so much the better as they become more diversified in structure, and are thus enabled to encroach on places occupied by other beings. Now let us see how this principle of great benefit being derived from divergence of character, combined with the principles of natural selection and of extinction, will tend to act.'

After the *Origin*, Lyell (1868, p.351), in the tenth edition of volume II of the *Principles,* had been officially and publically converted to natural selection by Darwin, and in doing so he misappropriated Matthew's principle of power of occupancy, which Selby (1842) had earlier cited and failed to understand. Typically, without citing the originator, Lyell passes it off as Wallace's discovery by referring to it as the "power of pre-occupancy," citing not Matthew's idea, but instead writing about Wallace's discovery that powerful sea currents kept species apart to explain it, exactly as Darwin (1859, p.403) did when he used it in the same way to explain pre-occupancy:

'...*pre-occupation has probably played an important part in checking the commingling of species under the same conditions of life. Thus, the south-east and south-west corners of Australia have nearly the same physical conditions, and are united by continuous land, yet they are inhabited by a vast number of distinct mammals, birds, and plants.*'

Going back to the first edition of that volume, Lyell comments without reference to the originator Matthew's key device of explaining natural selection by reference to artificial selection (Lyell 1832, p.26):

'*Now let us first inquire what positive facts can be adduced in the history of known species, to establish a great and permanent amount of change in the form, structure, or instinct of individuals descending from some common stock. The best authenticated examples of the extent to which species can be made to vary, may be looked for in the history of domesticated animals and cultivated plants. It usually happens that those species, both of the animal and vegetable kingdom, which have the greatest pliability of organization, those which are most capable of accommodating themselves to a great variety of new circumstances, are most serviceable to man. These only can be carried by him into different climates, and can have their properties or instincts variously diversified by differences of nourishment and habits. If the resources of a species be so limited, and its habits and faculties be of such a confined and local character, that it can only flourish in a few particular spots, it can rarely be of great utility.*'

But, as Eiseley (1979) remarked, Lyell initially veered away from recognizing the veracity of natural selection. Perhaps it was because Matthew's theory accommodated what we now know is the reality of geological catastrophes punctuating long steady-state periods. Lyell, the Christian, accepted organic evolution following the *Origin's* promotion of his erroneous uniformitarian belief that all past evolution occurred without catastrophe. Nevertheless, we can see that Matthew's example of the crab apple may have been playing on Lyell's mind. Only because he could see no evidence that two distinct species—the crab apple and pear—evolved from the quince. Lyell was convinced in 1832, the year following publication of *NTA*, that because no one had observed artificial selection producing new species, new species did not ramify one to another in nature (Lyell 1832, p.32):

'*They may be regarded as extreme cases brought about by human interference, and not as phenomena which indicate a capability of indefinite modification in the natural world.'*

As we know, and to necessarily repeat the point here, confirmation of evidence that Matthew's theory of natural selection, wedded to his politics, was too taboo to discuss in print, comes from a 20 page, otherwise highly favourable, review of *NTA*. The book review article spans Parts II and III (1831a and 1831b) of the *United Service Journal and Naval and Military Magazine*, which disclaimed Matthew's idea of the natural process of selection (1831a, p.457):

'In thus testifying our hearty approbation of the author, it is strictly in his capacity of a forest ranger, where he is original bold, and evidently experienced in all the arcana of the parentage, birth and education of trees. But we disclaim participation in his ruminations on the law of Nature, or on the outrages committed upon reason and justice by our burthens of hereditary nobility, entailed property, and insane enactments.'

Elsewhere, in the same year that Matthew so upset the establishment by wedding heretical science to seditious politics, a young Darwin, fresh out of Cambridge University and still believing in divine, miraculous, intelligent design creationism, set sail on the *Beagle*. Meanwhile in 1831, an anonymous author of the *Edinburgh Literary Journal* (1831, p.2) appeared also to be criticizing Matthew's politics and fully worked out scientific theory when they laid-into *NTA*:

'The entire tract resembles a new quack-medicine, full of high stimulant, ignorantly and not very safely combined, and which, till known and analyzed , might prove dangerous as well as attractive to young patients (i.e. young planters and country gentlemen), from the necessant puffing of the compounder.'

That seemingly mocking call for Matthew's ideas to be analyzed and known perhaps suggests that this particular anonymous critic wished to see the theory tested by induction (the bringing-in of examples) but could only say so safely by mocking the *Originator*. On which note, the above *Edinburgh Literary Review*'s (1831) reviewer quite correctly, describes *NTA* as divided into five parts, but the fifth part alone, presumably Note B of the Appendix, which contains a particularly condensed exposition of Matthew's theory, is neither named nor discussed in the review. The critic, who wrote only of four of the five parts they identified, merely quoted Matthew's own self-deprecating words, written at a time when his ideas on the topic were heretical, that on the topic of species he had *perhaps gone too wide.*

Identifying but then weirdly failing to address the fifth part of *NTA* may have been yet another case of an author feeling unqualified to comment on theological concerns, or else merely a slip-shod oversight. Perhaps it was for reasons of expediency that Matthew's theory was not covered?

NTA received several prominent reviews. Presumably, all the reviewers had to be cautious about citing heresy. After all, *NTA* shared the same London publisher as the journal that published the Edinburgh Literary Journal review. Following the bankruptcy of Archibald Constable and Co. in 1826, Matthew's London publisher Longman, Rees, Orme, Brown and Green took over *The Edinburgh Literary Journal.* In 1832, the journal folded, and in that same year Chambers, who had regularly contributed work to it, launched the *Chambers Edinburgh Journal,* and therein Robert Chambers cited *NTA* (Chambers and Chambers 1832).

Most importantly of all, it is also vital to remember another point, also already made. Namely, that in 1859, when Darwin sent Chambers a copy of the *Origin* to review, that review was apparently the first to second-publish Matthew's unique term for his discovery the "natural process of selection." We might have reason suspect, therefore, that Chambers had *NTA* at the very forefront of his mind when reviewing the *Origin.*

Whatever the reason for including Matthew's four-word phrase for his theory of natural selection in his 1860 review of Darwin's *Origin,* or perhaps it is just another of many multiple co-incidences, Chambers' usage of it appears to have escaped notice until detected with IDD in 2013 (see Sutton 2014a).

That Chambers first penned his own heretical book on evolution 12 years after he cited *NTA* is yet another newly discovered fact that changes everything currently believed to be true in the story of Matthew and Darwin, particularly regarding the significance of Matthew's newly discovered possible influence upon others who influenced Darwin and other naturalists, such as Wallace.

The fact that Darwin could convince the scientific community so successfully for over 154 years that Matthew's complete prior-published theory was merely scattered in pages of an appendix, which was not read by any naturalists, suggests that *NTA* was probably not well known among the living or active members of the scientific community pre-1860. If many people had published comments about

Patrick Matthew's theory, there would be no reason to write this book, because we would all know that Matthew was the discoverer of the process, originator of the theory, and had influenced Darwin to spend the best years of his life gathering evidence to support it. The fact that no one other than Matthew, by then aged 70, stood up in 1860 and argued with Darwin's excuses in the press, suggests that we should not expect to uncover a great many more published cases of *NTA* being cited before 1858. However, more probably do exist.

When Darwin (1861) claimed, in the third edition of the *Origin*, that he had never heard of Patrick Matthew's book until Matthew brought it to his attention in 1860, Loudon, Powell and Murray II were dead. Even if he had felt inclined to protest, Chambers was in no position to do so, since he was the anonymous author of the *Vestiges*. Murray III could scarcely protest either, since he was the publisher of Darwin's fallacious excuses for not having read Matthew's prior-published breakthrough.

Out of all those we know of who read *NTA*, and who were in some way socially connected to Darwin, the devoutly Christian Cuthbert Johnson, who was 62 years old in 1859, was in a position to protest about Darwin's excuses, assuming, of course, that he even read the third edition of the *Origin*. However, Johnson, perhaps above all others, would not want to break the rules of courteous science by going into print on the subject of Matthew's prior godless theory. Not just because he was apparently first to be second to borrow Matthew's term "adapted to prosper" without citing him, but then cited him in another publication (see Sutton 2014a), but more so for reasons of a religious nature. Most notably, after the publication of the *Origin*, there was fear among devout believers in the scientific community that the idea that science was the only truth was threatening faith in Christian revelations of truth. Johnson was one of the Royal Society signatories to the 1864-65 Scientists Declaration that, problematically, research into scientific truth was casting doubt upon the truth and authenticity of the Christian scriptures.

Selby was 73 when the third edition of the *Origin*, was published in 1861. It was the first version to carry Darwin's fallacious "no one read Matthew's ideas", excuses. Had he been aware of the existence of that false excuse, one cannot help wondering whether Selby would have been past caring about such issues.

Since the vast majority of people who read or hear the views of others do not provide future generations with their own mainstream published evidence of that fact, it seems reasonable to assert that, besides those who cited it, a great many more other people, including, perhaps, more naturalists known to Darwin, must have read *NTA*. Perhaps hundreds or even thousands of others did so before 1858, either through owning it at some point, borrowing it from friends, or else borrowing it from or reading it in a library.

Logically, if, after so many years of research and prolific networking, Darwin never heard of *NTA* from a single one of those who read and then cited it, he was a schnook and not a crook.

Far from being an unread book with an unread theory, the first systematic review of the literature with IDD reveals that for some 20 years *NTA* enjoyed some kind of international status among a generation of agriculturalists and naturalists.

At this juncture, it is pertinent to point out that although Darwin had a severely debilitating condition, which began to dominate his life from around 1838, he nonetheless posted and received over 15,000 letters of correspondence in his lifetime, with over 2,000 different people. His correspondents included hundreds of other leading scientists and thinkers. Over 9,000 of these letters are available on the excellent online Darwin Correspondence website, but thousands are missing.

Darwin had many opportunities, therefore, to learn about *NTA* from others and discuss it with them. Between 1842 and 1882, Darwin was out of the house for 2,000 days, and we know that he attended 16 meetings of the Council of the Royal Society (Colp 1977).

Why was Darwin unable to read the one book in the world he most needed to read, if other naturalists could read it, many of whom were in his close circle of friends? This question becomes doubly telling once we add the fact that so many other scientists at the time, including some who read and cited Matthew pre-1858, knew Darwin was working on the problem of species. Eiseley's (1959, pp.156-157) point is similar: *The development of the theory of natural selection is often dated casually from the time of the publication of the Origin of Species in 1859. Actually, its inception occurred far earlier than this date. Since Darwin discussed the subject with his intimates over a long period and it was rather widely known in professional circles that he was working on the 'species problem,' it is even difficult in some instances to know how far his influence extended before he published.'*

It seems highly unlikely that none of the seven newly discovered naturalists who read *NTA,* then cited it, and who knew the Hookers, Lyell, Huxley and Strickland, did not speak to any one of them about Matthew's conception. If we accept that rational premise, then, in turn, it seems highly likely that one or more of the Hookers, Lyell, Huxley and Strickland did read *NTA.* Anyway, regardless of whether they personally read it or not, they would surely have communicated second-hand knowledge, if they had it, of its bombshell breakthrough heretical contents, to Darwin at some point during the time he was working on replicating the exact same idea by means of nothing more than avidly reading the literature.

We should wonder what counter-evidence could possibly be found now to serve as proof that Darwin was right, rather than deceiving his readers, to knowingly lie that no naturalist/no one at all was aware of Matthew's ideas before he and Wallace replicated them.

Currently, de-facto fact denial by autonomous agents of the Darwin Industry appears to be the default tactic to address this economic and intellectual threat. For example, Knapton (2014), Telegraph (2014), Caven (2014), Lewis (2016), Alexander 2016, and Burdge (2017) have all reported on the "New Data". In her Telegraph Internet article on it, Science Editor of the Telegraph newspaper (Knapton 2014) reported the opinion of a historian of science, Professor James Moore:

'I would be extremely surprised if there was any new evidence that had not been already seen and interpreted in the opposite way.'

Moore, presumably having not examined the "New Data", questionably, effectively told the Telegraph's chief science editor that he was happy to reject the independently verifiable facts, whatever they were, so long as anyone interpreted the newly discovered and 100 per cent independently verifiable fact that Darwin's and Wallace's associates, influencers and facilitators did read and then cite *NTA* in the pre-1858 published literature, to mean they never really did!

On the same subject of those newly discovered facts in my peer-reviewed paper (Sutton 2015), Scottish journalist Michael Alexander (2016) interviewed another arch Darwinite historian of science:

'Dr John van Wyhe, a senior lecturer at the Department of Biological Sciences, at the National University of Singapore, said the recent claims by Dr Mike Sutton of... were "so silly" and 'based on such forced and contorted

imitations of historical method that no qualified historian could take it seriously.'

We should ask ourselves, therefore, whether the following original discoveries I made are silly. Is it silly that I totally, 100 per cent, disproved the claims of the most highly esteemed biologists Darwin, de Beer and Mayr, who all wrote that no one read Matthew's conception pre-1860? Is it silly that I did so by originally discovering that Selby cited *NTA* and then edited Wallace's (1855) Sarawak paper on evolution? Is it equally silly that I discovered Chambers' (1844) cited *NTA*. He being the very same Chambers who then wrote the hugely influential *Vestiges of Creation,* which so influenced both Darwin and Wallace? Is it silly that Darwin met chambers in the 1840's (Darwin 1847b) and corresponded thereafter? Is it silly that the highly networked Loudon (1831) cited *NTA*? Is it silly that William Hooker's regular correspondent, Jameson (1853), cited *NTA?* Is it silly that three other naturalists and several agriculturists cited *NTA* pre-1858? Is it silly then that in addition to all that, I originally discovered Loudon went on to edit and publish Blyth's (1835, 1836) influential articles, read by both Darwin and Wallace pre-1858, on adaptation within species? Is it silly that Darwin (see Darwin 1848a) met Blyth in person, read his work, and corresponded with him at least a decade before 1858? Is the associated explanatory concept of written and oral "knowledge contamination" (Sutton 2015) silly then, when I have proven Darwin lied about the pre-1858 readership of Matthew's breakthrough after his and Wallace's influencers and influencer's influencers cited *NTA* years before Darwin's and Wallace's claimed amazing independent replications of both the original theory in it and its essential explanatory analogy of differences? Is the alternative explanation of Darwin's and Wallace's independent immaculate miraculous conceptions of Matthew's prior published and cited theory, its unique explanatory highly idiosyncratic analogy of differences, and Darwin's use of the same four words to name it, not silly?

With respect, perhaps it would have been far less silly if Dr John van Wyhe had used Google before I did, and so found for himself what I found, before I found it for him, rather than launder in the Scottish press what might appear to some to be shrilly indecorous jealousy of the New Data. Will all such Darwinists remember to cite me as the originator of these newly discovered facts? Will they be considered "silly" scientific and historic facts then?

4 THE ROBERT CHAMBERS CONNECTION

The newly discovered fact that Robert Chambers read *NTA* is important in the story of the discovery of macroevolution by natural selection.

Many Darwinists consider Chambers an important precursor of the theory that no one was supposed to have read before Darwin and Wallace (1858) replicated it.

Millhauser (1959, p.84) saw Chambers as a great scientific thinker. He wrote of Chambers' (1844) 'Vestiges of Creation':

'One does not cut a scientific hypothesis out of the whole cloth. In a work like Vestiges (and this would hold for The Origin of Species too) the author's contribution is not the absolute and parthenogenetic conception of the idea. It is the recognition of its importance, its relevance and scope and probable validity; it is the sifting of materials, the marshalling of evidence, the construction of an argument; it is insight, selection, organization, interpretation.'

He is right in all of that. However, what Millhauser never knew when he dismissed any claim that Robert Chambers had prior solved the problem of species was that Chambers read *NTA* and then cited it. Chambers next cited Matthew's second book *Emigration Fields,* which took Matthew's ideas forward for humankind (Chambers 1840). Now, as we know, the world's leading expert biologists agree *NTA* contains the original complete theory of macroevolution by natural selection. Chambers was, therefore, surely aware of the fact that Matthew originated it. Furthermore, Millhauser was unaware also of another significant fact we now newly know. Namely, that in his book review of the *Origin,* Chambers (1859) inserted Matthew's apparently original term for the most important discovery ever made: "natural process of selection" as opposed to Darwin's (1859) multiply coincidental apparently original four-word re-shuffling of it to "process of natural selection". That is rather strong confirmatory circumstantial evidence that Matthew significantly influenced Chambers. In addition, we know Chambers influenced Darwin, Wallace and others, because Darwin and Wallace wrote with their own pens to admit as much.

Royal Society Darwin Medal winner Ernst Mayr claimed to be sure-fire robust in his criticisms. He wrote that he cared not to merely parrot the ideas of others in the field (see Cock and Forsdyke 2008). However, he did merely parrot the fallacious conclusion of others that Matthew influenced no one (Mayr 1982, p.500):

Patrick Matthew undoubtedly had the right idea, just like Darwin did on September 28, 1838, but he did not devote the next twenty years to converting it into a cogent theory of evolution. As a result it had no impact whatsoever.'

Mayr was wrong to voice such unevidenced beliefs, because today, thanks entirely to the newly discovered facts in *Nullius,* we know it is untrue to claim Matthew had no impact whatsoever. He appears to have had an impact on Chambers to write the influential *Vestiges.* However, we must also be clear that the evidence-led fact of the matter is we do not know *exactly* what impact Matthew had on those influential naturalists, whom we now newly know did read and cite his work pre-1858, before they influenced and /or facilitated the replicating work of Darwin and Wallace.

How many other discoverers of natural laws or solutions to great problems are we to strip of their status using Mayr's, "especially made for Matthew" criteria for greatness? The answer is that we could probably dispense with a great many famous scientific discoverers. For example, according to such conveniently fabricated criteria, we should promote Florey above Fleming for the discovery of penicillin as a systemic medicine, and the genius-level expert operatives of the CERN'S Large Hadron Collider above Englert and Higgs for the discovery of the Higgs Boson particle.

Shall we then use Mayr's no precise hard evidence of impact criteria to start retracting Nobel Prizes? Alternatively, should we instead publicize what these fact denying Darwinists have been up to all this time? Would we prefer to push Matthew back behind the dusty old curtains of Darwin's own stage? Should we pretend also that the Originator remains buried in oblivion in an unmarked grave? Should we not honour him by marking his re-discovered last resting place in Errol Churchyard with a memorial fitting for an immortal great discoverer and influencer in science? Alternatively, perhaps we should prefer the romance and lies about Darwin to the truth of his serial lying and plagiarizing glory theft. Perhaps we should never mention the names Matthew, Chambers, Selby, Loudon, Jameson and Wallace in the same book ever again. Shall we also conveniently ignore the typology of knowledge contamination?

Based as it is on newly debunked fallacies started by Darwin, Mayr's criteria for Matthew's greatness exclusion depends upon the fallacy that no naturalist who influenced Darwin or Wallace had read the original ideas in *NTA* before influencing them. As we know, that punctured premise fallacy has a long intellectual provenance going back to Darwin's essential and various "I never read it and neither did anyone else" proven liar excuse types.

Many years before Mayr made-up his greatness exclusion criteria to help keep Matthew buried in oblivion, Judd failed to take account of the fact that Matthew (1860, 1860b) explained to Darwin about Loudon citing his work. Judd also failed to factor-in the contextual account Matthew provided to Darwin about why and how his original work was brute censored for its heresy (Judd 1910, p.342):

'*...Matthew anticipated the views of Darwin on Natural Selection, but without producing any real influence on the course of biological thought...*'

Mayr had something of a habit of such bias. In relation to his dismissive and ill-informed treatment of another naturalist, Cock and Forsdyke (2008, p.623) wrote:

'*It is one thing to be buried and forgotten. It is another to be buried and have people come from far afield to stamp on your grave.*' That superb line fits like a glove Mayr's poor treatment of Matthew.

To understand more of the truth about Matthew's newly discovered role in informing evolutionary biological thought, and the historic role of Darwinists in thwarting intellectual objective fact-led enquiry into Matthew's probable influence on it, we can see that Wallace had a big hand to play in this traditional fact denial and context free history telling. Wallace (1871, pp. iv-v) wrote:

'*It therefore happens, that, while some writers give me more credit than I deserve, others may very naturally class me with Dr. Wells and Mr. Patrick Matthew, who, as Mr. Darwin has shown in the historical sketch given in the 4th and 5th Editions of the 'Origin of Species,' certainly propounded the fundamental principle of 'natural selection' before himself, but who made no further use of that principle, and failed to see its wide and immensely important applications.'*

Interestingly, Wallace, who, unlike Darwin, never had to tell an outright lie in his replicator defence, does not explicitly write, as Darwin did, that he never read Matthew's theory, nor does he claim explicitly that no other naturalist read it. Wallace appears, however, to be the prior published source of Dawkins's (2010) the "poor sucker, never knew what he had" rationalization for denying Matthew's greatness for the discovery, invention and creative explanation of natural selection. One can only wonder what Wallace would have had to say about the fact that Gregor Mendel, likewise, made no further use of his great discovery of the laws of inheritance, which only years later informed the study of genetics.

In the realm of scholarly research, Chambers (1859) provides one clue as to what he may have been up to in his book review of the *Origin*:

'It will be interesting to observe the effect, in the scientific world, of such views brought forward on scientific grounds by a naturalist of eminence.'

This is a most informative sentence, because Chambers uses the choice phrases, "views brought forward," "on scientific grounds" and "naturalist of eminence." Arguably, these all imply that Darwin did not originate the views in question. Chambers appears to be making it very clearly known that those views are merely brought to the fore by Darwin, and that they are not, therefore, his own discovery. Chambers wrote that Darwin brings views forward on *scientific* grounds. I perhaps go too far by asking whether that is a nod to what he believes to be Matthew's non-scientific deduction of the original theory. Similarly, by writing that Darwin is a "naturalist of eminence," would it be a leap too far to ask whether he might just have been alluding to the fact that Matthew is not?

Staying with mere speculation for a moment longer, for all we know, Matthew's godless explanation for the origin of species, used to justify libertarian politics, might well have played some influential role in shaping Chambers' political shift from Tory leanings towards his later appreciation that the solution to science problems lay in hard science, not religious rhetoric. *NTA* may have influenced him also to weave the evolutionary philosophy of both inorganic and organic development throughout the *Vestiges*. However, arguably, the *Vestiges* contains no shred of Matthew's macroevolution by natural selection theory. Perhaps the reason for its absence was Chambers' anger at Matthew for mocking his personal hero and sponsor, Sir Henry Steuart. Alternatively, perhaps it was because he considered Matthew's original and complex ideas on organic evolution to be interesting, but equally unfounded and muddleheaded. The fact is, we just do not know. And so it is best to see such speculation for what it is, admit as much and place the utmost value on independently verifiable facts only. There are plenty enough of those in this book.

Notwithstanding the typical silent treatment meted out to any scientific publications transgressing into the realm of religious revelations, the *Vestiges* became a best seller, escaping oblivion due to Chambers' popular writing style, marketing genius and skilful dedication to promoting his book. Consequently, while never having a single declared adherent (Barnes et al. 1996), it did attract in due course some exceptional published venomous criticisms from a small handful of writers. Perhaps because silence failed, they hoped to bury it under informed ridicule. However, that failed in the end, because we know that Darwin and Wallace read the *Vestiges*, and both wrote that they saw its importance and influence. By way of example, in his personal notebook of books to read and books read, which he began in 1838, Darwin wrote:

'Vestiges of Nat: Hist: of Creation. Churchill: 1844. 7s 6d in which species are shown to be not immutable see Brit. Museum Collect. (Anonymous).'

In addition, in his historical sketch in the third edition of the *Origin*, Darwin (1861) argued against some of the ideas in the *Vestiges*, regarding how species might evolve by leaps, but fully admitted their value:

'The work, from its powerful and brilliant style, though displaying in the earlier editions little accurate knowledge and a great want of scientific caution, immediately had a very wide circulation. In my opinion it has done excellent service in calling in this country attention to the subject, in removing prejudice, and in thus preparing the ground for the reception of analogous views.'

Wallace's 1845 letter to his fellow specimen hunter, Bates (see Shermer 2002), is sufficient proof of fact of the impact of the *Vestiges* on *his* thinking and on his plans to collect empirical evidence to support the general theory of evolution of varieties and species:

'I have rather a more favourable opinion of the Vestiges than you appear to have. I do not consider it hasty generalization, but rather as an ingenious

hypothesis strongly supported by some striking facts and analogies, but which remains to be proved by more facts and the additional light which more research may throw upon the problem.'

Shermer sought to explain how thinking about Chambers' *Vestiges* influenced Wallace's thinking. However, it now looks probable, in light of the "New Data" that Matthew's thinking directly influenced Chambers' research and thinking, which then influenced Wallace's and Darwin's research and thinking. This knowledge about "knowledge contamination" routes (Sutton 2015) stems from the newly discovered fact, unknown to Shermer (2002), that it was Matthew who, albeit to some unknown degree, probably influenced Chambers' thinking on organic evolution. Unwittingly, therefore, in light of what we now newly know about that, Shermer in fact documents the probable extent of Matthew's hidden influence upon Chambers (1832, 1844) and, in "domino effect" consequence, all other naturalists who read the *Vestiges* before 1858, including Darwin. Most ironically, Shermer (2002) writes:

This 'species question'—what is the difference between a variety and a species, and if a variety varies enough from its original type, can it become a new species?—had a long pedigree that Wallace would inherit and take with him to the tropics. Clearly Vestiges had an impact on Wallace, since he immediately began speculating on the relationship between geography and change within both varieties and species. In fact, he became an evolutionist shortly after reading Vestiges, and shortly before heading for South America on his first voyage.'

Shermer (2002) fallaciously implied that Matthew believed in fixity of species. More on that below, but first it is important to recognize that Shermer is not the first to make that particular ill-informed "mistake". For example, in January of its first year in print Matthew's publishers placed at least two large advertisements for *NTA* in the *Quarterly Literary Advertiser* (1831); one in January, and another in November. The adverts, while prudently avoiding any mention of evolution, had quite a lot to say about Matthew's views on species and varieties:

'In embracing the Philosophy of Plants, the interesting subject of Species and Variety is considered—the principle of the natural Location of Vegetables is distinctly shown—the principle, also, which, in the untouched wild, 'keeps unsteady Nature to her law' inducing conformity in species, and preventing deterioration of breed, is explained, and the causes of the variation and deterioration of cultivated Forest Trees pointed out.'

This subject matter of the advert, if it caught the eye as intended, would most certainly have been of interest to any economic botanist at the time. Furthermore, and surprisingly by today's figures, *NTA* was one of only five botanical books published in Britain in 1831 that was not authored by William Hooker, father of Darwin's best friend. What is more, it was reviewed and boldly advertised as such alongside Hooker's own work in *The Gardener's Magazine* in 1832. Hooker, it seems reasonable to assert, would almost certainly have read the advert placed by Matthew's publishers.

Unsurprisingly, in light of what we know about the context of times, that advertisement, like so many others for *NTA* (see Sutton 2014a), did not share the details of Matthew's heresy about evolution leading to the origin of new species. An explanation for the misleading advertisement can be taken, once again, from information provided by Secord's (2000, p.64) scholarship regarding prejudicial attitudes towards all deductive works on natural history at the time:

'A causal account of generation of higher species was, even in the most liberal medical circles, simply too speculative to be an available subject for general treatise, regular research, or a lecture course. Instead 'unity of type' was the battle cry of the earlier reformers, not the origin of new species.'

We can see now why Shermer had no need to worry about the historical facts concerning how 19th century scientists responded to heresy when he disappointingly duplicated the fallacy that Matthew's version of natural selection was limited to fixity of species, "species preservation." By this device, the evolutionary "expert" Shermer facilitates Darwin's project to bury the threat of Matthew in oblivion. Shermer then, contrary to what the world's leading biologists have written on the matter, continues by spreading the dysology that Darwin and Wallace alone came up with the full theoretical scientific solution to the problem of species (Shermer 2002, pp.147-148):

Darwin and Wallace, among their peers, synthesized a vast quantity of biological and geological phenomena in parallel fashion different from what anyone else had done. But most of the bits and pieces were already there. What they did with these intellectual parcels is what is original to them. Matthew, Blyth or others (e.g., William Charles Wells, discussed as another 'precursor') may have predated Darwin and Wallace with a similar idea, but they did nothing with it. Wallace, and especially Darwin, took this mechanism of species preservation and changed it into one of species transmutation, then constructed a research program to test the theory, and in the process took a giant leap foreword in our understanding of the origin of species.'

One can only assume that perhaps Shermer was not quite skeptical enough to bother reading Matthew at his 1831 source.

Another Darwinist, Kentwood Wells (1973) made a more scholarly attempt at explaining away *NTA*. Having at least, apparently, read *NTA* before writing about it, Wells claimed, wrongly, that Matthew's origination of natural selection explained species evolution occurring, but only after geological extinction events. Wells thought that in the periods between extinction events, Matthew merely saw natural selection as simply preserving species by way of a hypothetical process, allowing only the best circumstance suited individuals to survive, thus non-circumstance suited mutations never thrived in competition with those that were already best circumstance suited. Critical of this misrepresentation of what Matthew had plainly written, Dempster (1996, p.161) wrote that Matthew's acceptance of the role of catastrophes in evolution horrified Wells because it clashed against Darwin's rejection of the idea. Most interestingly, to repeat the point already made, a fact ignored by biased Darwinists such as Wells, is that Matthew's emphasis on geological catastrophic extinction events has turned out to be correct (Rampino 2011). Most importantly, the point is that Wells got it completely wrong when he claimed that Matthew thought species only originated from the need to adapt to startling changes in circumstance following catastrophes. Wells (1968, p.250) wrote:

'Furthermore, it is clear that Matthew did not see natural selection as a mechanism leading to the extinction of species.'

In fact, Wells provides us with a perfect model example of mythmaking. He intertwines Matthew's text with his own, therein smogging the fact that Matthew's version of natural selection was completely different from Darwin's. Here is Wells' (1968 p.247) subversion of Matthew's version:

'...periods of rapid evolutionary change were followed by 'millions of ages' of stability, during which no further evolution occurred.'

The point is that "...during which no further evolution occurred," are the words of Wells not Matthew! Therefore, Wells fabricated a falsehood by mixing his own words in with those of Matthew, which makes it appear as though Matthew penned the entire sentiment. Wells does this by way of enclosing his own words, not Matthew's, within inverted commas. For the sake of veracity, here are the six relevant words that Matthew actually penned about natural selection, which clearly highlight Wells' and Shermer's unique fallacy spreading (1831, p. 385): *"This principle is in constant action."*

While Matthew did think, apparently, that catastrophes were the most important reasons for extinctions, his work reveals that he was fully aware that competition between species could cause extinction of the losers, which is what he called a "natural process of selection." After all, Matthew was, as noted by Dempster, living in an age where man had already made extinct in Britain the bear, beaver and wolf. Unlike Darwin, Matthew saw human beings as inseparable from culture and society in natural history. Humans were right at the centre of Matthew's work on evolution. He focused heavily on this highly complex evolutionary theme, both in *NTA* and in his later book *Emigration Fields*.

Matthew (1831, p. 387) wrote on this topic:

'As far back as history reaches, man has already had considerable influence, and made encroachments upon his fellow denizens, probably occasioning the destruction of many species, and the production and continuation of a number of varieties or even species, which he found more suited to supply his wants...'

That last paragraph alone busts the myth so cleverly started by Darwin that Matthew was a simple-minded Noah's flood biblical catastrophist.

Kentwood Wells (1968) gets the facts completely wrong yet again where he anticipates Shermer's "fixity of species" mistake (p.249):

'Since Matthew did not believe in continuous evolution, it is obvious that the action of natural selection must have been independent of evolution during the long interludes between periods of change.'

Whereas in his overview of *NTA*, Dempster (1996, p.150) almost correctly outlines exactly what Matthew's notion of natural selection was:

'Matthew introduced the concept of Natural Selection as a fundamental law of nature, in addition he discussed divergence in terms of diverging ramifications, the mutability of species, rejected miraculous birth or new species following catastrophes, held to a steady state in nature interrupted by catastrophes, rejected development from nearly-allied species in favour of descent from common ancestor, recognized the difference between domestic and wild species, and recognized what constituted a species.'

However, Dempster made a mistake by omission, because his

own work reveals he knew Matthew also attributed some species extinction to the natural process of selection in the relatively stable periods between catastrophes. I stand on Dempster's shoulders to write that Matthew's true origination of macroevolution by natural selection is as follows:

Matthew originated the concept of natural selection in 1831 to explain the emergence and extinction of species *between* and *after* geological catastrophic events. He uniquely named it the "natural process of selection," which he described as a fundamental law of nature. He discussed divergence in terms of diverging ramifications. He accepted the mutability of species. He rejected miraculous birth of new species following catastrophes. He held to a general, but not absolute, steady state in nature interrupted by catastrophes. He understood the importance of the complex multi-level phenomenon of power of occupancy and ecological niches. He rejected simple development from nearly allied species in favour of descent from a common ancestor. He recognized what constituted a species. He recognized the difference between domestic and wild varieties and saw artificial selection as the powerful explanatory analogy of differences key to both discovering and explaining the process of natural selection, which he coined a "natural process of selection".

Whilst the amateur naturalist Darwin speculated that any slightly profitable variation in varieties of species would become the norm, Matthew, a commercial hybridizer of fruit trees, avoided this (see Dempster 1996, pp.151-152), because practical experience taught him that slight and profitable variations would not cause species change. For his part, Darwin wrote nothing of catastrophes influencing species change. Dempster's (1996, p.202) criticism of what he sees as Dawkins' misreading of Darwin on this issue is particularly enlightening. Contrary to modern day Darwinist spin, Punctuated Equilibrium Theory is, arguably, in fact Matthewism and not Darwinism. Well, at least it is according to Dempster.

There are three other contenders for the title as originator of the theory of natural selection. They are Hutton (1794), Wells (1818) and Blyth (1836). However, all of Matthew's pre-1858 rivals are easy to eliminate. Starting in date order with Hutton, here is why.

Hutton understood the adaptive changes that organic matter undergoes to survive in particular circumstances. His published work suggests he believed in the fixity of species, as does Buffon's.

Moreover, he adopted Buffon's definition of species, in that he explained that what definitely separates and so defines one species from another is when one type cannot breed with another. In Hutton's published notion of organic evolution, species do not diverge and ramify in this way beyond the original (Hutton 1794, p.499):

'In like manner, when a plant or animal is produced by the propagation of the species, the individual is never precisely the same as that which had preceded it; but, while it thus varies according to contingent circumstances from the parent, it does not transgress the order observed in the species. So far, therefore, as the nature of things admits, the species may be changed in continuing the race, that so it may be always properly adapted for the purpose of its existence, in a world where varying circumstances would require a certain difference of constitution for the individuals, who have to find their sustenance amidst extreme difficulties, occasioned by a changing state in the circumstances of their life and manners.'

What about Wells? William Charles Wells, who studied medicine at Edinburgh University between 1775 and 1778, and anatomy under William Hunter (George 1996), espoused some natural selection-type ideas about human skin tone variation. His ideas were limited to within species variation and focused almost entirely on humans. With regard to animals, he touched very briefly upon variation only (see Calman 1912, Eiseley 1959), no mention was made of the plant kingdom. Wells says nothing of the extension of change by way of unrestricted variation in unlimited time through the species barrier. Most importantly, Wells makes no mention of extinction of species. Wells' essay is perhaps not as original as some claim, because, as Eiseley (1959) points out, in his opinion it is in fact no more original than an earlier, 1786, paper by Townsend.

Essentially, William Wells merely speculated, with reference to animal breeders, that the skin of a settlement of white people would have to evolve and become black if they wished to survive in certain regions of Africa (Wells 1818, pp.435-436):

'Again, those who attend to the improvement of domestic animals, when they find individuals possessing, in a greater degree than common, the qualities they desire, couple a male and female of these together, then take the best of their offspring as a new stock and in this way proceed, till they approach as near the point in view, as the nature of things will permit. But what is done in art, seems to be done, with equal efficiency, though more slowly, by nature, in the formation of varieties of mankind, fitted for the country which they inhabit. Of the accidental

varieties of man, which would occur among the first few and scattered inhabitants of the middle regions of Africa, some one would be better fitted than others to bear the diseases of the country. This race would consequently multiply, while the others would decrease, not only from their inability to sustain the attacks of disease, but from their incapacity of contending with their more vigorous neighbours.'

Mayr (1982, p.499), who wrote, "The person who has the soundest claim for priority in establishing a theory of evolution by natural selection is Patrick Matthew," also highlights the limitations of Wells' paper:

'Although Wells clearly proposes a theory of evolution by natural selection, it is only evolution of adaptation to local climates within a species and at that for man only; the principle is never applied to genuine evolution, to the multiplication of species, to a development of higher taxa or to common descent.'

We should not discount the mere possibility that Wells' essay influenced Matthew and then Darwin. Matthew used the word "vigour" in the same sense that Wells used "vigorous" to describe his perception of the general constitutions of black Africans.

In a letter that Darwin (1865) wrote to Hooker, he admits that Wells has priority over himself for natural selection for that 1818 paper on skin pigmentation. Darwin writes snidely:

'So poor old Patrick Matthew is not the first, and he cannot or ought not any longer put on his Title pages the Discoverer of the principle of Natural Selection!'

Here we see Darwin wriggling to conceal the true originator. In this particular instance, by claiming Wells has priority over Matthew, Darwin rationalizes a way to deny Matthew's origination of the key importance of divergence and ramification in explaining extinction and common ancestry shared by distinct species.

Darwin stressed the importance of divergence and ramification of species in the *Origin*. If Darwin was not well aware of the fact that his own unique work on divergence and ramification merely confirmed Matthew's prior-published theory and right to true discoverer status, he was in a state of denial.

According to de Beer (1960), Charles Lyell and Edward Blyth both had some pre-*Origin* priority for natural selection, but they saw it as evidence that evolution could not occur. de Beer claims that Wells and Matthew also had prior-publication priority over Darwin, but then de Beer credulously accepted and regurgitated the Darwinian excuse myth that they both failed to appreciate the significance of their observations and failed also to provide any evidence to support the theory or work out its consequences. In that regard, he was right only about Wells. It is true that both Wells and Matthew did not provide anything like as many evidences as Darwin. However, Matthew never intended to provide a multitude of evidences, his obvious motive being to originate the theory so that others— naturalists such as Darwin—would come along and discover the necessarily detailed evidences for it, or else to refute it. As he wrote (Matthew 1831, p.386):

'In the first place, we ought to investigate its dependency upon the preceding links of the particular chain of life, variety being often merely types or approximations of former parentage; thence the variation of the family, as well as of the individual, must be embraced by our experiments.'

The fact of the matter is that Matthew should have been officially awarded full and clear first and foremost scientific priority over both Darwin and Wallace, because he published the discovery of natural selection first and is proven to have influenced influential naturalists and others to cite and comment on it before Darwin's and Wallace's replications. Furthermore, the Arago Case norms of priority, as affirmed by the Royal Society in the case of Le Verrier's discovery of Neptune in 1846, that, in the absence of plagiarism, being first into published print with an original discovery settles all questions of priority for both first and foremost discovery attribution (see Biagioli 2012, Strivens 2003 for more details). Therefore, at least according to official science norms and conventions, Matthew has both first and foremost priority over Darwin and Wallace. Nevertheless, for some unknown reason or reasons, or perhaps for no reason other than ignorance, the so-called "scientific establishment" and its leading scientists have effectively concealed that one particular fact from us.

As Strivens (2003) explains: *'Science's priority rule rewards those who are first to make a discovery, at the expense of all other scientists working towards the same goal, no matter how close they may be to making their discovery.'* Robert Merton (1957) famously explained that this strongly entrenched convention applies even if one person is beat by a matter of hours or even minutes. That fact alone makes the Darwinist excuse myth mugging of Matthew all the more despicable, because both Darwin and Wallace were not even working on the same problem when Matthew published his findings in 1831, both still believing species to be immutable.

When Darwin did realize that species could be explained by

evolution, Chambers' anonymous authorship of the *Vestiges* was a great secret, because such an idea was heretical. As early as 1847, Chambers was an acquaintance and correspondent of Darwin, and had actually given him a copy of his heretical book.

Amongst many accused of authoring the *Vestiges,* both Darwin and Lyell were fingered as suspects (Secord 2000, p.21). In the last lines of a letter, Darwin (1847) wrote to Hooker about his personal association and correspondence with Chambers he lets it be known who the secret author is:

'I think I have only made one new acquaintance of late, that is R. Chambers, and I have just received a presentation copy of the 6th Edit of the Vestiges: somehow I now feel perfectly convinced he is the Author. He is in France & has written to me thence.'

A decade later, Darwin either genuinely feared, or else feigned fear, that Chambers might plagiarize him. In which vein, he wrote to Asa Gray, bidding him to keep the correspondence about his unpublished essay on natural selection secret, lest it fall into the hands of the author of the *Vestiges* (Darwin 1857):

'You will, perhaps, think it paltry in me, when I ask you not to mention my doctrine; the reason is, if anyone, like the Author of the Vestiges, were to hear of them, he might easily work them in, & then I shd' have to quote from a work perhaps despised by naturalists & this would greatly injure any chance of my views being received by those alone whose opinion I value.'

Darwin considered the *Vestiges* a work despised by naturalists. He wrote to Gray as though that was a matter of agreed fact, which provides insight that *NTA* would most likely have been similarly despised by other naturalists, given that it broke so many conventions. Why did Chambers give Darwin a copy of the *Vestiges*? This question has its twin in the question: Why did Wallace first write to Darwin, of all people, about his work on the origin of species? The answer to this "why did they give them to Darwin?" question is most likely that both Wallace and Chambers knew that Darwin was working on the problem of species, it surely being no secret among the markedly clubbable, conference, committee and correspondence loving, intricately networked gentlemen of science. William Hooker, for example, could have told Wallace, since they were regular correspondents, personally acquainted and because Wallace made no secret of his own interest on the exact same topic.

We know, from the dates he wrote on his personal private

notebooks that Darwin claimed he began looking at transmutation of species from 1837. It seems that Chambers probably shared an interest in this subject from at least around the same time. Yet current knowledge, informed only by what survives of Darwin's correspondence, has it that Darwin sought out a meeting with Chambers primarily to discuss what was to became a rather famous debunking of Darwin's own erroneous theory regarding Glen Roy having been under the ocean.

Whatever the main reason was for them getting together, Darwin wrote to Chambers on February 28, 1847, and they first met the following Wednesday. Initially, Chambers sided with Darwin's conclusion that the ancient marine beaches of Glen Roy were a geological feature of the area, but later sided with the more conclusive evidence that a freshwater lake was the cause. In a letter to Hooker (Darwin 1847a), Darwin confessed that the existence of evidence that he may have been wrong made him sick on two occasions:

I have had several long letters to write lately on Glen Roy, which has vexed me much—Mr Milne has been trying to prove the former existence of common lakes, which I feel sure is absurd, but his paper staggered me in favour of Agassiz ice-lake theory, so I wrote a letter to the Scotsman. Now R. Chambers, who was a follower of me, & then became a convert to Milne, has been there again, & now says he can prove the sea theory—The confounded subject has made me sick twice.'

Yet another "multiple coincidence" in our story is that before Darwin and Wallace pondered the same subject area, Matthew (1831) speculated about large swathes of Scotland having been under the ocean. So here, once again, we have Darwin's interests colliding with those of the man who he portrayed at turns as merely an obscure Scottish writer on naval architecture, or else forest trees.

These multiple coincidences of Darwin following in Matthew's footsteps just keep mounting up. And so it is possibly no more than one more mere coincidence that *NTA* featured Matthew's own ideas regarding the geological history the Carse of Gowrie having once been the bottom of a lake. Matthew (1831, p.378) wrote:

This carse appears to have been a general deposition at the bottom of a lake having only a narrow outlet communicating with the sea, and probably did not rise much higher than the height of the bottom of the outlet at that time...'

"Coincidentally", in 1848, sixteen years after he cited *NTA* in his own journal (Chambers and Chambers 1832), we find Robert Chambers published a book entitled *Ancient Sea Margins as Memorials of Changes in the Relative Level of Sea and Land* (Chambers 1848). Moreover, he presented and published no less that 11 learned papers on this theme between 1850 and 1857. One was on an ancient boat hook found in Matthew's immediate neighbourhood of the Carse of Gowrie (see Millhauser 1959, p.214). His papers on geology include one given at Cambridge University, attended by Darwin's mentor, Charles Lyell, who then wrote to his father, from his Scottish manor house, just 20 miles from Matthew's manor house, that he knew Chambers was the anonymous author of the *Vestiges* (Lyell 2010).

This research proves that Matthew and Darwin were much closer together in their wider influences than previously known. We can see now that Darwin not only read Chambers's *Vestiges*, and was presented with a copy by its author, he was influenced by it. And, lest we forget, Darwin was given that copy of the 6th edition of the *Vestiges* by Chambers – a naturalist who read *NTA*, cited it, and then used the most important original phrase from it in his review of the *Origin* years later. That is confirmatory evidence of Matthew's likely "knowledge contamination" influence on Chambers, which is most important because Chambers is well known to have influenced Darwin, Wallace, Powell and other naturalists. To necessarily repeat the point already made, the myth that Matthew failed to influence anyone with his discovery is rebutted by this new evidence of routes for oral or written "Matthewian knowledge contamination" of the pre-1858 minds of Darwin and Wallace, via Chambers.

Keen Darwinist, David Leff's website, AboutDarwin.com, is a useful resource on this topic. Regarding those who most significantly influenced Darwin, Leff writes about Robert Chambers:

'In October 1844 Chambers published (anonymously) a controversial book titled, 'Vestiges of the Natural History of Creation.' This was the book that brought the notion of transmutation out into the public arena. It attempted to describe the entire evolution of the universe, from planets to people, as being driven by a self developing force which acted according to natural laws. The book was written more for the poor working class of England rather than the scientific elite for it appealed to their desire to 'evolve' beyond their wretched economic circumstances. The book received widespread criticism, mainly because the ideas it contained went against the old scientific school which adhered to the idea that nature did not evolve according to unguided laws, but rather by the divine hand of

god. Despite the harsh criticism, Vestiges sold very well.'

Chambers' noteworthy influence on Darwin is expressed in the Historical Sketch in the third edition of the *Origin* (Darwin 1861) and every edition thereafter. Therefore, to necessarily really hammer home the point already made, if the fact that he cited it (Chambers 1832) supports the probability that Matthew influenced Chambers' *Vestiges,* which we know influenced Darwin and Wallace, we have evidence for Matthewian "knowledge contamination" of the pre-1858 brains of Darwin and Wallace, a line of influencer-provenance going back to the man who influenced the man who influenced the man.

Moving on, here is another "coincidence" for the multiple coincidences collection in this story: Matthew apparently coined the exact phrase "living aggregates" in *NTA* (Sutton 2014a). Within three years of that publication, the natural theologian and physician Peter Mark Roget (1834, p.591), of *Roget's Thesaurus* fame, was using Matthew's phrase in his publications on the subject of evolution of species.

Coincidentally, Roget, the staunch Church of England member, is noted to have once failed to cite, despite replicating, the important and original work of another fellow scientist. He left the Royal Society under a cloud in 1848 following years of systematically blocking the work of the Nottingham born physician Marshall Hall (Manuel 1996, p.157), who, like Matthew, was a man of action, rather than a full-time, privileged and non-employed Christian gentleman naturalist. But that's not all. It seems Roget, like Darwin, was a serial plagiarism offender because he is famously known to have plagiarized Robert Grant's discoveries (Desmond 1989a, p.234). This is the very same Grant who made the mistake of confiding in Charles Darwin about his important new breakthrough on sea sponges (Desmond 1989a), which then motivated Darwin to dart off, frantically gather extra evidence for Grant's otherwise unknown discovery and then present it at a learned society in order to grab some self-glory for an idea he was incapable of discovering himself.

In the same publication where he plagiarized Grant (Roget 1834), we now know that Roget used Matthew's apparently original phrase. Moreover, as Desmond (1989a) points out, we should not forget that it was Grant's upbraiding of Darwin's academic encroachment, without due reference to the provenance of *his* unique discoveries, that led to Darwin falling out with his Edinburgh mentor and, in turn, may have sparked Darwin's complex reluctance to accord due priority to any of his future influencers.

Dempster (2005) explains that Grant introduced Darwin to the ideas of Lamarck and Cuvier while Darwin was a student at Edinburgh University, and yet Darwin never once acknowledged his tutor's influence. We can end this small line of our inquiry with the same kind of story-ending symmetry beloved of mythmakers by weaving in what Dempster (2005, p.103), for his part, has to say about the relationship between Darwin, Grant and two other unacknowledged influencers of the *Origin:*

'Grant, Blyth and Matthew were kept [by Darwin] at a distance. It is a very curious fact that these three all died within six months of one another, and from then on Darwin's psychosomatic illness faded and he enjoyed better health than in the previous 40 years.'

If that's not simply one more coincidence, then it is one among many possible explanations for why Darwin was so ill and what appeared to cure him. But that's all we can say about it. Writing it does not turn such a remote possibility into a measurable probability.

As I earlier wrote, the coincidences regarding *NTA,* just keep on coming. Here comes another, of which we might objectively make more, since it strengthens a social network connection between Darwin and those who read *NTA* and why. Roget and Robert Chambers were fellow members of the Geological Society of London, along with Darwin's close friends Lyell and Huxley. Notably, it is breakthroughs in geological understanding of the great age of Earth that turned over 2000 years of speculation on organic evolution into a scientific explanation (e.g. see Howard 1982).

Yet another coincidence is that Chambers was particularly interested in arboriculture, of all things. He and his brother published an extensive 15 page guide on the subject (Chambers and Chambers 1842), just two years before he first published the *Vestiges.*

One year after the publication of *NTA,* in 1832, Chambers's journal published a special volume on the life of Sir Walter Scott— the Scottish hero who was both ridiculed and criticized in *NTA.*

Sir Walter Scott was Chambers's patron (Secord 2000, p.82). The Chambers' volume also contained a two-page article on trees and arboriculture, based almost entirely on praising the work of Sir Henry Steuart, whose views Matthew also had the temerity to mock mercilessly in *NTA*.

Most remarkable of the many coincidences between the content of *NTA* and Chambers's rich life is the fact that in the early 1830s, Chambers was a church-going Tory and anti-transmutationist, which means that Matthew's godless treatise, riddled through with reformist politics, would have very much set Matthew in the mould as one of "the enemy" on all three fronts. But, having read *NTA* in 1831, before the decade was out, we know Chambers changed his politics to support for the liberal Whigs. He then ceased attending church, firmly took to believing that religion had no place in science and supported evidence for transmutation. Although, like Darwin (1861) we have seen also that he safely wrote into his work the notion that "The Creator" divinely designed things so that evolution would happen.

Was Chambers' radical transformation in some way caused by his coming into contact with Matthew's heretical and socio-political theory? Either it was simply another mere coincidence that he read it and so changed, or else Matthew, the Chartist, atheist evolutionist, had some kind of subversive impact upon Chambers's original conservative politics, leading him to reverse his political, religious and scientific views and then go into anonymous print on the subject of organic evolution.

Chambers was fully hexadactyl, meaning he was born with six-fingers and six toes Was that condition one more component part of the unique combination of forces that led Chambers to write the *Vestiges,* a book with the ultimate conclusion that everything, including humans, was evolving?

Did Chambers at times resent the surgical removal of his mutant fingers and the removal of his extra toes that left him lame? How could he, a man who pulled himself up by his own bootstraps to establish a mighty publishing empire, a man who once stood for political office, not at times consider himself a superior adaptation to the circumstances of the modern world, a mutant variety to thrive in the future? From his own words we might think we see a clue that Chambers believed he and his brother were evidence of humans evolving into a higher order of humanity (Chambers 1845, p.207):

'Is our race but the initial of the grand crowning type? Are there yet to be species superior to us in organization, purer in feeling, more powerful in device and act, and who shall take a rule over us? There is in this nothing improbable on

other grounds. The present race, rude and impulsive as it is, is perhaps the best adapted to the present state of things in the world; but the external world goes through slow and gradual changes, which may leave it in time a much serener field of existence. There may then be occasion for a nobler type of humanity, which shall complete the zoological circle on this planet and realize some of the dreams of the purest spirits of the present race.'

According to the leading expert on the subject, the *Vestiges* was widely accepted in its time as the one book that all readers of the *Origin* had read (Secord 2000, p.39).

On the subject of mere coincidence, just because "vestiges of creation" rhymes with Matthew's (1831, p.287) apparent coining of the phrase "vestiges of aeration" proves nothing. And so it would be preposterous, wouldn't it, to claim on this evidence alone that Chambers was making some in-joke at Matthew's expense?

Moving on to further fanciful uncertainties, it is surely yet another mere coincidence that Darwin, Chambers and Wallace all followed-up Matthew's recommendations to study the meteorology of the ocean. For example, Darwin and Chambers both visited and wrote, respectively, wrongly and rightly, about the influence of the sea, or else fresh water, upon the geological features of Glen Roy (Darwin 1839, Chambers 1848). And Wallace determined the impact of geological change and the ocean's currents on species distribution in order to discover his famous Wallace Line. The wonderful coincidence being that Matthew (1831, pp.246-247) wrote:

The vestiges of olden time, the exuviae of former worlds, in the exposed strata the abrasion of the rocky land by the continued battering of the numberless pebbles moved backward and forward by the heaving of the ceaseless wave Let them study the currents and winds and meteorology on the ocean.'

As if that many coincidences are not funny enough, the improbable multiple coincidence pestilence becomes almost too exquisitely unbearable with the inclusion of Matthew's (1831, p.246) career advice for all young naturalists, two of whom went on to steal his one really big idea:

Let their ideas shoot while they recline under the lone magnificence of the primeval forest while they gallop over the unappropriated desert, free of the Bedouin.

Let them learn geology and mineralogy on the Andes and Himalaya, and around every shore where the strata are denuded. Let them wind about among those abrupt rocks and craggy precipices, where they may contemplate the sea bird's household economy—the wild herbs of the cliff—the vegetation and shells and monsters of the ocean.'

We know that Wallace slept many a night in tropical forests, shooting rare birds and orang-utan mothers with infants, as well as ideas. We know also that Darwin, coincidentally, explored the Andes and the Cordillera desert on horseback (Darwin 1839a) for official geological purposes—most likely "free" of Bedouin. We know too that Darwin, coincidentally, contemplated plenty of shells, herbs, cliffs and ocean vegetation.

As for Darwin's encounters with sea birds, among such abrupt rocks and craggy precipices, much has been written. But far less well known is the tale of Darwin's geological hammer being hurled to slaughter trusting creatures. Perhaps there were indeed sea monsters for such a typically egocentric youth as Darwin to contemplate while he was enjoying his hooting slaughter. Darwin may have been killing just for meat, but the Beagle's Captain's log perhaps suggests otherwise. Captain FitzRoy (1839) wrote:

'When our party had effected a landing through the surf, and had a moment's leisure to look about them, they were astonished at the multitudes of birds which covered the rocks, and absolutely darkened the sky. Mr Darwin afterwards said, that till then he had never believed the stories of men knocking down birds with sticks; but there they might be kicked, before they would move out of the way.

The first impulse of our invaders of this bird covered rock, was to lay about them like schoolboys; even the geological hammer at last became a missile. 'Lend me the hammer?' asked one. 'No, no,' replied the owner, 'you'll break the handle;' but hardly had he said so, when, overcome by the novelty of the scene, and the example of those around him, away went the hammer, with all the force of his own right-arm.

While our party were scrambling over the rock, a determined struggle was going on in the water, between the boats' crews and sharks. Numbers of fine fish, like the groupars (or garoupas) of the Bermuda Islands, bit eagerly at baited hooks put overboard by the men; but as soon as a fish was caught, a rush of voracious sharks was made at him, and notwithstanding blows of oars and boat hooks, the ravenous monsters could not be deterred from seizing and taking away more than half the fish that were hooked.'

Here's another coincidence. Two of Matthew's sons, who possessed at least one copy of *NTA* between them, became friends in New Zealand with Darwin and Joseph Hooker's regular correspondent Sir George Grey, who served as governor of New

Zealand and was, most coincidentally, godfather to Patrick Matthew's grandson Duncan Matthew (see Tee 1984). According to Jones (2000), Patrick Matthew's sons supplied Governor Grey's estate with plants from their nursery sometime during his second term of office (1861-1868). Grey was a correspondent of both Darwin and Joseph Hooker. However, it is clear that he only met Matthew's sons after the *Origin* had been first published. This is just one more fascinating twist in the interconnected lives of Matthew, Darwin and Wallace, as seen through the histographic record of *NTA*.

In the *Vestiges* (1844, p.65), in the same year that Darwin penned his second unpublished essay on natural selection, Chambers employed, but did not articulate, a basic premise of Matthew's natural selection theory. Namely, he implied that species evolve by being selected by nature to be circumstance-suited, and within such prevailing circumstances conform to a degree of uniformity of suitable species, so consequent lack of variety will, for periods of time, necessarily ensue. Chambers borrowed that notion to explain that in the past, when more of the globe had a tropical climate than today, there were similar tropical species all over the globe. Unsurprisingly, given his interest in the subject, at pages 83 and 84 of the first edition of the *Vestiges* (1844), we find Chambers writing more on *NTA* subject matter, this time about the evolution, succession and modern day predominance of current species of British forest trees, no less!

In the preface of his tenth edition of the *Vestiges* (Chambers 1853), six years after he met with Darwin and gave him a copy, Chambers anonymously self-celebrated his bestselling publication. By that stage he had gone much further than in the first edition. The *Vestiges* now included a multitude of examples to support arguments for the transmutation of species by a simple process of linear development:

'As is well known, the fate of the book was not to rest in obscurity or oblivion, but to be extensively read, and become the subject of much animadversion. It has never had a single declared adherent—and nine editions have been sold. Obloquy has been poured upon the nameless author from a score of sources—and his leading idea, in a subdued form, finds its way into books of science, and gives a direction to research. Professing adversaries write books in imitation of his, and, with the benefit of a few concessions to prejudice, contrive to obtain the favour denied to him. It is needless to say that the storm of opposition has never for a moment affected his original faith in the hypothesis—as how, indeed, could it,

when not one of the writers on that side proved himself to have taken up a correct conception of the aim of the work, showed a power of reasoning upon it logically, or seemed capable of taking a candid view of the data, on which it rests?'

As we know, in 1857, so afraid was Darwin of Chambers' progress in proving the natural selection hypothesis that he asked Asa Gray not to discuss his own ideas with anyone else, unless the author of the *Vestiges* included them in his next edition. In reality, Chambers was, arguably, little more than a compiler who was unable to correct many of his own scientific errors in earlier editions of the book without the help of experts (Secord 2000).

As we have seen, in additional to the cultural codes of the day, Chambers had many personal reasons for wishing to see *NTA* buried in oblivion. Having once proved that he read *NTA* by citing it on pruning (Chambers 1832), he was perhaps ultra cautious thereafter to avoid including its unique evolutionary ideas in his own precious *Vestiges,* even if that would have meant passing them off as his own in order to avoid bringing *NTA* to the attention of a wider public.

Chambers knew only too well of the published rumours about him being the author of the *Vestiges,* since, as we have seen, they thwarted his earlier political ambitions. And he knew that once he died, his authorship would be declared.

While it might be easier, more memorable, marketable and crowd pleasing to write a conveniently simple linear account of Matthew's influence, the convoluted reality of life is sometimes more complex. As Professor Iain McCalman of the University of Sydney explains, for example, Wallace was influenced by Chambers' *Vestiges* to write his Sarawak paper. And yet the *Vestiges* influenced Darwin to initially misinterpret it (McCalman 2009):

'On a first reading of Wallace's paper, Darwin himself had detected nothing new. 'It seems all creation with him,' he scribbled in the margin. Darwin thought the paper to be simply another version of the thesis that had been expounded by an anonymous bestseller Vestiges of Natural Creation in 1844 that attracted much scientific odium when it was published. Darwin was right in guessing the source of Wallace's evolutionary ideas. When Wallace read the Vestiges, it had been the single most important intellectual experience of his life. The book had been written by Robert Chambers, a talented Scottish journalist of radical and free-thinking views who managed to blend wild speculations with an innovative natural history of the earth. As the title hinted, the book rejected a literal version of the biblical creation in favour of a materialist interpretation of the origin of the firmaments,

organic life on earth and of global geological change. Chambers carefully hid his subversion. He gave the divine creative responsibility for setting these natural laws in progress. He didn't use the term 'evolution' but he had sketched out an historical theory of the biological interconnection of species and of what he said 'the progress of organic life upon the globe,' and it implied an evolutionary law.'

~~~

Dear reader, at this juncture I thank you for your perseverance in reading so far, and beg your tolerance of the several necessary repetitions you have endured in reaching this point in *Nullius*.

I think that the newly discovered data has brought us a very long way from the currently accepted romantic rhetoric that Matthew's theory could not have had any reasonably probable impact whatsoever on Darwin. We know, by way of just one example, among many we now have of his newly proven impact on other naturalists, which was enough for them to read his *NTA* book and then cite it, that Matthew must have had an impact on Chambers, who we know had an impact on Wallace and Darwin. Do we prefer our knowledge evolves with significant and newly discovered, independently verifiable, facts like this one? Or do we prefer to regurgitate and swallow debunked romantic rhetoric, proven lies and the jealous dishonesty of the Darwin Industry?

The facts are that Chambers cited Matthew in 1832. He then met with Darwin in the decade before Darwin penned the *Origin*. A route for both oral and written Matthewian knowledge contamination of Darwin's brain is proven by facts such as these alone.

5 EARLIER INVESTIGATIONS

During 22 years of published expert scholarship, comparing Darwin's work with Matthew's, Dempster (1983, 1996 and 2005) proves, with reference to the historical and written record, that Matthew fully understood the importance of his breakthrough. He shows how Matthew took it forward in *Emigration Fields*. What is more, he shows that it was superior to Darwin's replication. In addition, he shows how Darwin cleverly dumbed-down Matthew's importance.

Revealing in painstaking detail the full extent of Matthew's original conception of macroevolution by natural selection, Dempster makes an extremely good scientific case for Matthew's complete priority over Darwin. However, those carefully balanced arguments failed to influence leading Darwinists, because Darwinists ignored the knowledgeable detail Dempster uniquely provided and simply retorted in Darwin's defence that Matthew failed to influence any naturalists. Until now, nobody produced any disconfirming evidence for that myth, other than Loudon's (1832) review, which Darwin scholars, generally, cannily choose to ignore, or else know nothing of its existence. However, to repeat for the benefit of any fact-blind cult of Darwin worshippers, the point made several times in this book is that we now newly know for a fact that Darwin's excuse is a busted myth.

Dempster's fact-driven reasoning that Matthew should be duly recognized and celebrated as an immortal great of science was thwarted by the unevidenced rhetoric of leading Darwinists, such as Mayr, Gould, Shermer, Hamilton and, most recently, Dawkins. All their skilfully conveyed yet ill-informed rhetoric depended upon Darwin's now debunked, self-serving myths and lies about Matthew's original ideas being unread by naturalists before 1858.

Rather hit and miss in their research of the question of Darwin's plagiarism, Eiseley and Grote (1959) created etymological mistakes in some of their evidences, although IDD confirms they were right about some. The problem is, without the benefit of today's Big Data technology, they were unable to demonstrate which phrases were apparently genuinely unique to Matthew and which were definitely published before Darwin replicated them.

Milton Wainwright (e.g., 2008, 2011) of Sheffield University has written several articles on the theme of Darwin plagiarizing Matthew's breakthrough. He is convinced of Darwin's research fraud based on the sheer weight of collective circumstantial evidence he has compiled and discovered, along with some reasonable suppositions and the application of common sense. For example, he finds it highly unlikely that the Beagle would not have had on board a copy of *NTA*. He was also the first to spot that *NTA* was reviewed alongside Lyell's and Lindley's books (Loudon 1832), a fact that he, quite reasonably, believes would surely have led Darwin to see Loudon's review.

Wainwright also doubts Darwin would have missed prominent advertisements for any book being on the subject of species and variety, particularly one quoting Matthew on the subject of the origin of species. Wainwright (2011, p.16) writes:

'Clearly anyone, including Darwin who was interested in the 'species question' would have read this and wondered what this somewhat elusive quote meant.'

Importantly, Wainwright (2011) makes it abundantly clear that both Darwin and Wallace fully acknowledged that Matthew completely got there before them with the prior published discovery of macroevolution by natural selection. He points out, essentially, that Kentwood Wells' (1973) Darwinist argument that their ideas were different is patent nonsense. Wainwright demonstrates that it is based on Wells' ludicrously arrogant and delusional premise that neither Darwin nor Wallace understood natural selection as well as he understood their more limited than his understanding of it.

The philosopher Hugh Dower's website, HughDower.com, has a page (Dower 2009) that adopts the same weight of circumstantial evidence approach taken by Wainwright. On the site, Dower adds more examples of advertisements for *NTA* existing in pre-*Origin* literature. In addition, he thinks it is possible that, following Matthew's 1860 letter in the *Gardeners' Chronicle*, Darwin ordered and received a copy of Matthew's book in four days. However, he thinks it highly unlikely he could have done so. Dower thinks it far more likely that Darwin always had his own copy. Moreover, Dower informs us of his finding that on the day Matthew's letter appeared in the *Chronicle*, both Hooker and Huxley were spending the weekend at Darwin's house. Judging by the tone of Darwin's letter to Hooker, asking him to approve, re date, and then send his reply on to the *Chronicle*, it does appear they probably had a prior discussion regarding what to do about the Matthew matter.

Dower (2009) also surmises, although hard evidence is non-existent for this suspicion, that pages removed from Darwin's notebooks were most likely about trees. Dower believes the removal of those pages was because of their links between his work and Matthew's. Back on solid evidence-led ground, however, Dower notes Darwin's intense interest in the subject of varieties of oak trees, which was one important topic of *NTA*. Dower, like Eiseley (1979) before him, notes also the similarities between Darwin's unpublished essay of 1844 and *NTA*, on the subject of Matthew's highly idiosyncratic original artificial versus natural process of selection explanatory analogy of differences between trees grown in nurseries compared to those in the wild. And of course, we know also that Wallace (1858) used the general analogy of differences between natural and artificial selection in his Ternate paper, and that Darwin used that same more general analogy to open Chapter One of the *Origin*.

Eiseley's (1979) original discovery of this most suspicious replication of Matthew's idiosyncratic forestry and arboriculture, essential explanatory analogy of differences with nature, is one that is, to date, universally cannily ignored by Darwinists. Of course they don't, but it is almost as though they have a little rule book somewhere forbidding mention of it. Darwin's and Wallace's replication of this analogy is either a shocking discovery of evidence for plagiarizing science fraud, or else just one more incredible aspect of multiple coincidence, in light of what we now newly know about Matthew's actual pre-1858 readership and what Darwin wrote about artificial selection in the biography edited by his son (Darwin 1887, p.83):

'My first note-book was opened in July 1837. I worked on true Baconian principles, and without any theory collected facts on a wholesale scale, more especially with respect to domesticated productions, by printed enquiries, by conversation with skilful breeders and gardeners, and by extensive reading. When I see the list of books of all kinds which I read and abstracted, including whole series of Journals and Transactions, I am surprised at my industry. I soon perceived that selection was the keystone of man's success in making useful races of animals and plants. But how selection could be applied to organisms living in a state of nature remained for some time a mystery to me.'

Without the benefit of IDD, Dower used a search engine to conduct an etymological investigation of Darwin's use of the word "plastic," on grounds that it was used in *NTA*. He writes on his website (Dower 2009):

'What Eiseley seems to have missed is that, in the famous Appendix, Matthew refers to the 'plastic quality of superior life', and that Darwin frequently describes organised life as 'plastic' in both the 1842 sketch and the 1844 Essay. That may seem insignificant, but I have put search engines through numerous natural history texts of the period on the internet (including Herbert's 'Amaryllidaceae' and Chambers' 'Vestiges....'), and have found no other use of the word plastic.'

Here we see that without the power of IDD to search over 35 million publications, using words, terms and phrases as evidence of plagiarism leads scholars to fall into that old trap of the etymological fallacy. Without IDD, Dower could not in several lifetimes of sweaty little orange rubber thimble page turning and other mind-numbing library research have the remotest chance of making such a claim with a shred of validity. IDD, however, proves that he is wrong to rely on use of the word plasticity as evidence of Darwin plagiarizing Matthew. The fact of the matter is that the word "plasticity" and "plastic" exists abundantly in print before. By way of just two examples among a multitude, Taylor (1665) uses it. Here is a quote from Cudworth and Birch (1820, p.387):

'That besides that plastic principle in particular animals, forming them as so many little worlds, there is a general plastic nature in the whole corporeal universe, which likewise, according to Aristotle, is either a part and lower power of a conscious mundane soul, or else something depending on it.'

The same word exists pre-*Origin* in a publication noting the superiority of the wild ass over the domestic, which *NTA* may have possibly influenced (Partington 1838, pp.221-222):

'Plastic animals which break into varieties adapted to different climates and modes of treatment may be improved by culture, so that the domesticated shall, in the qualities which are desired; be much better than the wild, but those which like the ass, not so plastic, cannot be improved or even kept up to their natural state if domesticated. For this reason the domestic ass is in all countries inferior to the wild ass; and though the differences of those of warm and cold climates be very considerable, they are only indications of different degrees of deterioration.'

Moving on from Dower's unintentional fallacy spreading to his better work, the question of the likelihood that Darwin became aware of *NTA* via Loudon's 1832 review is capable of examination by what Dower reveals in Darwin's unpublished notebooks. Here Dower unearths some of the most compelling examples of circumstantial evidence in this story. Like life itself, it is all far too intricately involved and entangled to summarize in this book, so I suggest you visit his website (Dower 2009) to see and judge it for yourself.

While some might determine that Wainwright and Dower's conclusions suggest a case for more in-depth research, their conclusions are not of themselves sufficient to prove Darwin's pre-1860 plagiarizing science fraud. Nevertheless, I cannot leave it at that, because the critical scholarship of these authors involved incredibly painstaking research, the kind that is seldom recognized or rewarded, and is likely to lead to future unwarranted Darwinist rejection and ridicule. The fact is that their findings add to a considerable collective weight of evidence supporting the new fact-based evidence presented in this book that Darwin and Wallace more likely than not committed science fraud by plagiarizing glory theft pre-1858.

Clarke's (1984) conclusion that Darwin read *NTA* but forgot, depends solely upon his intuitive perception of the implausibility that anyone as well networked and thirsty for facts as Darwin could have missed it. That sensible speculation quite rightly failed to convince anyone that common sense is sufficient evidence. However, it does identify a problem that Darwinists have chosen to ignore. Namely, the "dual independent conception problem". Darwinists have studiously avoided researching it. The question it raises is *"how could Wallace and Darwin plausibly have avoided reading the book that contained the very theory they both replicated, the same four words to name it and the same analogy of differences to explain it, when others they knew well, who influenced them both, read it?"*

Today strong evidence to support one answer is that it seems improbable they could have missed it, because all around them too many other naturalists did read it before 1858. Moreover, many of those who read it were their personal associates and influencers, or their influencer's influencers, who knew of their profound general interests in organic evolution. This newly unearthed data supports Dempster's earlier rationale for granting Matthew foremost recognition over Darwin and Wallace as an immortal great thinker and influencer in science.

The main point of this chapter is to fulfil the customary requirement of clearly demarcating the contribution of my research from that of my predecessors. Therefore, I am obliged now to further reveal and emphasize yet more of the relative limitations and respective errors that IDD has enabled me to uncover. By so doing, I am empowered to use these results by dint of the scholarship of the very giants upon whose shoulders I now stand to criticize—it being their earlier wisdom and results of their painstaking work that informed, influenced and guided me from the outset.

In his 1996 book entitled *Patrick Matthew and Natural Selection*, Dempster seemingly implies an unacknowledged influence of Matthew's work on Wallace's. He points out the well-known fact that in 1855, Wallace sent a copy of his published Sarawak essay to Darwin. In looking at the problem of descent, Wallace's essay contained the phrase "closely allied species." This was a notion that Matthew had rejected in favour of descent from a common ancestor, the exact same conclusion that Darwinists believe Darwin arrived at independently.

Dempster reveals that Wallace never cited the fact that, 19 years earlier, Blyth (1836) used the phrase "species nearly allied." For his part, Darwin (1859) fails to cite them both on this topic, and yet, with great frequency, he used the phrase "closely allied species" to explain the law of natural selection (e.g., Darwin 1859, p.478):

The existence of closely allied or representative species in any two areas, implies, on the theory of descent with modification, that the same parents formerly inhabited both areas; and we almost invariably find that wherever many closely allied species inhabit two areas, some identical species common to both still exist. Wherever many closely allied yet distinct species occur, many doubtful forms and varieties of the same species likewise occur. It is a rule of high generality that the inhabitants of each area are related to the inhabitants of the nearest source whence immigrants might have been derived. We see this in nearly all the plants and animals of the Galapagos archipelago, of Juan Fernandez, and of the other American islands being related in the most striking manner to the plants and animals of the neighbouring American mainland; and those of the Cape de Verde archipelago and other African islands to the African mainland. It must be admitted that these facts receive no explanation on the theory of creation.'

With regard to these apparently multiply amazing etymological coincidences, Dempster (1996) is rather obtuse. However, with careful scrutiny, it is possible to distil that on this one point he is in fact informing us:

Matthew (1831) writes of "species nearly allied."
Blyth (1836) to writes of "species nearly allied."
Wallace (1855) writes of "closely allied species."
Darwin (1859) to writes of "closely allied species."

On the basis of these etymological clues, Dempster, like Eiseley (1859), concludes that both Wallace and Darwin took ideas from Blyth's work published between 1835 and 1837, but failed to reference any of these particular articles. Crucially, as Dempster (1996, pp.187-188) points out with clarity, the phrase "species nearly allied" was not coined by Blyth. In fact, Matthew (1831, p.384) used the exact term five years prior to Blyth. On this particular observation, Dempster writes:

'I presume Matthew picked the problem up in Paris because I cannot find it discussed in the English literature.'

Dempster's discussion, as always, studiously falls short of brash accusations of plagiarism. However, left hanging as it is, he strongly implies a story of shameless plagiarizing where Blyth plagiarized Matthew, and Wallace plagiarized Blyth, followed, presumably, by Darwin copying from Wallace.

Of course, we now know yet another most important newly discovered fact, which once again builds on earlier important findings. This particular one is that Blyth's editor and publisher of those two articles was Loudon, who had earlier (Loudon 1832) written that Matthew apparently had something original to say "on the origin of species". So here, we now have not only the replicated term but also the route for knowledge contamination by which it may have travelled to Blyth.

Like Dempster, we cannot prove Blyth read *NTA*. However, unlike Dempster we have now proven that Blyth's editor and publisher read *NTA* before Blyth replicated that particular term that is in *NTA*. Nevertheless, this means very little if the term was not unique to Matthew.

Fortunately, IDD allows us to cut straight through this unsatisfactory, farcical speculation to see whether the phrases "species nearly allied" and "closely allied species," are in the literature before Matthew supposedly first coined them. They are.

Had the technology and IDD expertise been available to Dempster in 1996, he could have conducted an Internet Big Data analysis, as opposed to what would otherwise have required an international army of thousands of researchers to plod through every word of the most vaguely relevant printed literature in order to avoid his own etymological mistake. Dempster's mistake is proven, because IDD proves that the term "nearly allied" appears in relation to species at least as far back as the late 17th century, where Molyneux (1695, p.180) writes:

'*...we observe that Nature affects the like disparity in other of her Works, and those too nearly allied, and evidently of the same Tribe or Family.*'

In the late 18th century (Heister 1750, p.220), we get even closer with the use of "species is nearly allied." Soon after, the exact phrase, "species nearly allied," (Lightfoot 1777, p.949) is occurs during the description and classification of a leaf:

'*The immersion of the seeds in the substance of the leaf makes this species nearly allied to the genus of ULVA; but being collected into warty clusters, it seems to be joined more naturally to the tribe of FUCUS.*'

After the end of the 18th century, the phrase occurs many times in the relevant naturalist literature before either Wallace or Darwin used it.

Traditional, expert, scholarly library research then, no matter how dogged, cannot compete with even a non-expert informed, researcher deploying IDD. Suffice it to say, the implications of this are rather startling.

Another example of the pre-IDD limitation of relying upon words and phrases to suggest plagiarism is that the accuser could not possibly know, without asking them, from where the accused author might have got what appears to be a rarely used word or phrase. For example, Dempster (1996, p.85) thought that Matthew was first to use the term "selection" in a philosophical sense. Not only is that proven untrue with IDD, the method allows us to prove that the exact term "natural selection" was used in a philosophical sense by Corbaux (1829) even earlier. This particular important fact, Milton Wainwright found first (See Wainwrightsceince.com), presumably using a similar Boolean Google search method to IDD that I employed later.

Chapter Eighteen of the extended e-book of *Nullius* (Sutton 2014a) reveals that apparently only four people published the exact phrase "natural selection" pre-*Origin*. None of these four was a naturalist, agriculturalist or breeder. And since we know already that Darwin claimed he got the phrase from unremembered literature on breeding (see Darwin's March 30, 1859 letter to Lyell), the closest match out of the millions of books and articles currently available online for IDD is Matthew, who did write about breeding and hybridizing trees. Moreover, to emphasize the point once again, it is his uniquely essential phrase "natural process of selection," which we know Darwin four-word-shuffled in the *Origin* to "process of natural selection." IDD enabled me to investigate and untangle things in a way that Dempster never could.

In his apparent attempt to keep Matthew buried in oblivion with one-sided, Darwin-friendly inquiry, Gould (2002) essentially wheeled out the Mayr Myth to accuse Eiseley of committing what he called an "etymological mistake." Gould claimed that, *"Natural selection ranked as a standard item in biological discourse."* The implication Gould made is that it cannot have been coined by way of a direct line of influence from Matthew's absolutely essential phrase "natural process of selection." Despite providing zero evidence to support it, Gould's winning argument is accepted by Darwinists. Accepted by them as proof that Eiseley was naively mistaken in thinking "natural selection" was a rare term. In fact, IDD proves Gould wrong. He was desperately "bullshitting" in the philosophical sense generally described by Frankfurt (2005). As Chapter Eighteen of the *Nullius* e-book (Sutton 2014a) reveals, pre-*Origin* publications incorporating the term "natural selection," or anything remotely close to it, were, apparently, extremely rare indeed.

Wilson and Kelling's (1982) criminological "Broken Windows Theory" explains that initial signs of neighbourhood incivility create an escalating spiral of decline by signalling lack of guardianship to offenders. Analogously, letting scholars get away with publishing fallacies and myths signals to others the existence of topics where guardians of good scholarship might be less capable than elsewhere. Such dysology then serves as an allurement to poor scholars to disseminate existing myths and fallacies, and to create and publish their own. This is the Dysology Hypothesis.

Once one side succeeds with fallacy spreading, as the Darwinists have, they create a niche where they have the power of occupancy to dominate all around them and repel invaders, because intellectual guardianship of veracity is relatively incapable.

On the subject of fallacy spreading, by someone with no such current dominant power of occupancy on this topic, Dempster claimed that Matthew coined the phrase "diverging ramifications." However, once again IDD reveals a case of fallacy mongering on his part, because that phrase was earlier published in a book on natural philosophy in 1790 (Nicholson 1790, p.309). What is more, it occurs many times in other works in the late 18th century, and particularly, for some reason, those related to the history of Jacobism (e.g. Barruel 1798).

Interestingly, Hope (1831) published in the same year as *NTA* his essay entitled *The Origins and Prospects of Man*, where he too used the phrase "diverging ramifications" on page 48.

Hope's explanation for the observable differences between black Africans, white Europeans and Malay peoples is that each different so-called "original type" of human was created separately and divinely, as were all species. That reveals how much further and uniquely advanced were Matthew's original ideas than those of his Regency contemporaries, who were trying also to understand the problem of variety and species in all organic lifeforms.

In seeking out data on influences, priority and levels of understanding between Matthew, Blyth, Wallace and Darwin, Dempster got his facts wrong and missed an important clue regarding Wallace's replication of Matthew's breakthrough. Dempster (1996, p.217) wrote, "Matthew did not use 'type' in his essay although he was acquainted with Cuvieran ideas."

This is wrong. Matthew does in fact use the term twice in the context of natural selection.

On page 386 of *NTA,* he writes:

'This continuation of family type, not broken by casual particular aberration, is mental as well as corporeal, and is exemplified in many of the dispositions or instincts of particular races of men.'

And again, on page 371:

The changes which have been taking place in France, and which, in many places, leave now scarcely a trace of the fine race which existed twenty centuries ago, may however, in part, be accounted for by the admixture of the Caucasian and Keltic tending more to the character of the latter, from the latter being a purer and more fixed variety, and nearer the original type or medium standard of man…'

Davies (2008, p.2) in his book, which focuses almost exclusively on bold claims that Darwin plagiarized the work of Alfred Wallace, observes that in 1856, Wallace wrote to Darwin from the Malay Archipelago. Although that particular letter is another that is frustratingly missing, we know of its existence because Darwin's reply is in the public domain.

Davies highlights the fact that Darwin's letter of reply explained the importance of four things, all of which are keys to explaining the origin of species, which Wallace had been working on trying to crack. Up to that point in time, Davies claims Darwin had neither published nor written in his private journals about them. They are (a) divergence, (b) modification, (c) extinction and (d) divergence, linked to extinction. The major point made in this part of Davies' book is that all four concepts are central in Darwin's *Origin*.

On this premise, Davies believes Darwin solved the problem of the origin of species by plagiarizing Wallace's ideas.

However, the fact of the matter is that Davies appears to have zero knowledge of *NTA,* because divergence, modification with dissent and extinction, once explained in relation to slow natural selection extinction as a process, leaping mutation, geological extinction events, competitive struggle for survival, multi-level circumstance-suited power of occupancy and ultimately impermanent ecological niches, are all uniquely Matthewist evolution by natural selection concepts. Matthew alone first melded these key Lamarckian concepts into a full and complex, part intuitive, part evidence-led counter-intuitive, explanatory theory for the origin of species.

Some of the individual components of Matthew's theory were most likely influenced by the work of Buffon and Cuvier, as well as others, all without due citation. Most significantly, Matthew published that incredible breakthrough 25 years earlier than Wallace did!

In sum, Davies's book, far from proving that Darwin plagiarized

Wallace, actually incriminates Wallace for plagiarizing Matthew's breakthrough. Davies should have read *NTA*. If he never, then Davies should have then read Dempster's priceless scholarship on the topic. Had he done so he would have found the centrepiece of the puzzle of Darwin's supposed Immaculate Conception of the same breakthrough, which was supposedly dually immaculately conceived by Wallace; or else explained rationally by Matthewian knowledge contamination via the many routes for it newly discovered to have existed before 1858.

Much is in the literature on the topic of Darwin's questionable originality focuses on Blyth's un-cited influence upon Darwin's *Origin*. However, as with the case of Wallace's influence, this is a red herring. Because Darwin knew that Blyth, too, wrote nothing fundamental on natural selection that Matthew had not beaten him to in writing on the process and validating it with his explanatory analogy of differences.

Davies (2008, p.27) cites from Blyth's (1835) paper, '*Just as man is able to affect the physical constitution and adaptations of domestic animals, so wild nature might achieve the same success,*' as evidence that Blyth influenced Darwin. However, in the book containing Matthew's (1831) prior-published theory, four years earlier than Blyth's efforts, those exact same ideas were explicitly available in print. In fact, that one simple idea was abundant in the literature in the early 19th century before Matthew wrote on the subject. What Matthew (1831) uniquely did that Darwin (1844) replicated in a private essay and then to open Chapter One of the *Origin* was to apply the same explanatory analogy of differences to artificial selection and natural selection. Matthew, the forester and Scottish writer on trees had due cause to use trees to demonstrate that very example. Is it just yet another among so many amazing multiple coincidences, that from the comfort of his study at home, Darwin (1844) replicated Matthew's highly idiosyncratic forestry example as well?

From his own synthesis of the literature, Davies, who in the 1980s made the famous BBC television documentary "The Devil's Chaplain," concluded (Davies 2008, p.162):

Now I am convinced that Charles Darwin—British national hero, hailed as the greatest naturalist the world has ever known, the originator of one of the greatest ideas of the nineteenth century—lied, cheated and plagiarised in order to be recognised as the man who discovered the theory of evolution.'

Whatever their discoveries and shortcomings, arguments that Darwin arrived at the theory of natural selection by plagiarizing Wallace (e.g., Brackman 1980, Davies 2008) or Blyth (e.g., Eiseley 1979, Barrett et al. 1987, Davies 2008) are now essentially relegated. This is because Matthew's discovery was published years before Blyth or Wallace ever put pen to paper on the same subject and, of course, Matthew's breakthrough was cited by Loudon (1832), who we know went on to be editor of two of Blyth's most influential papers, and by Selby, who went on to edit Wallace's Sarawak paper. Consequently, Matthewian knowledge contamination of Wallace's pre-1858 brain via Blyth, who's 1855 article Wallace read (Costa 2014) and via Selby, is arguably more probable.

Every book and paper arguing that Darwin plagiarized Blyth can be summarily dismissed and fairly because Darwin did cite Blyth's later work and did admit, in the third edition of the *Origin*, the great general debt that he owed to Blyth.

To our knowledge, Blyth never once complained or challenged Darwin. There is, therefore, no news story to tell about Darwin plagiarizing Blyth other than that Darwin's failure to cite all the examples originally published by Blyth, which supported Matthew's prior discovery, read and cited by Blyth's editor, Loudon. Of more concern to scientists and historians of science should be the fact that Blyth, Wallace and Darwin all failed to cite Matthew's prior published discovery, which some of their fellow naturalists, such as Loudon, Selby, Chambers and Jameson, did manage to read and cite.

To excuse Darwin's excuses for not reading Matthew's book, Darwinists including Dawkins (2010) and Hamilton (2001) have sought to convince us that forest trees and fruit trees were obscure areas of interest that were irrelevant to Darwin's interests. In reality, we have seen that independently verifiable science facts trump mere unscientific biased belief rhetoric every time, because Darwin's (1844) private essay, Darwin's notebooks of books read and the annotated books in his library reveal he was obsessed with trees including fruit trees (see Sutton 2014a for full details) as a means to explain natural selection. The first words of Darwin's Zoonomia notebook (Darwin 1837-38) are on the topic. Darwin and Hooker corresponded regularly about trees. Most remarkably, if *NTA* really was an obscure book, the telling question is: 'why then do we newly know that Darwin's notebook of books read (Darwin 1838) reveals that he held in his hands at least five publications that cited it (Athenaeum 1839, Loudon 1831, Loudon 1838, Gardener's Magazine 1841, Memoirs of the Caledonian Horticultural Society of Edinburgh 1814-1832)? Or is that just another set of amazingly mere multiple coincidences?

6 DISCUSSION AND CONCLUSIONS

In the 19th century, even to this day, many people do not like the idea that humans and chimpanzees descended from a now extinct species of ape. A common humorous jibe on social media is simply to recite the ignorant argument of creationists "If humans evolved from monkeys [apes], why are there still monkeys?" Today, many Darwinists don't like the idea that the *Origin* and *Vestiges* descended from the same common ancestor: *NTA*. Perhaps some will soon ask: "If the *Origin* evolved from *NTA*, why is there still *NTA*?"

If one book evolves from another and proves to be more successful than its ancestor, a succession of transitions enforced by a succession of some advantages, theoretical, or evidential and cultural, seems probable.

As confirmatory evidence that the *Origin* descended from *NTA*, Darwin's (1842, 1844) private essays, Wallace's (1855) article, Darwin's and Wallace's (1858) papers, and Darwin's *Origin* (1859) collectively replicate Matthew's (1831) prior-published original conception of the theory of macroevolution by natural selection, his unique and essential explanatory analogy of differences, same the four words he used to name it, three of them being central, and many other unique creative examples that he used as evidence.

The New Data completely disproves Charles Darwin's (1860a, 1861, 1861a) widely believed story about there being a complete absence of scientific readership of Patrick Matthew's (1831) original discovery of the process of macroevolution by natural selection before he and Wallace replicated it in 1858 without citing the originator. These and all the other newly discovered facts that you have read in this book come from Big Data technology unearthing long neglected but incredibly important printed matter in the historic publication record. The New Data is confirmatory evidence for earlier suspicions (e.g. Eiseley 1979) that Darwin stole the entire concept of natural selection from Matthew before passing it off as his own. Moreover, it points a new verifiable fact-led finger of suspicion at Alfred Wallace for doing the same.

Society has absorbed as true an escalating number of fake facts ever since Darwin replied to Matthew's (1860) priority assertion letter in the Gardeners' Chronicle. To address those proven falsehoods with veracity, the following points summarize 10 key independently verifiable fact groups covered in this book:

1. Darwin essentially admitted that Matthew got the entire theory of macroevolution by natural selection before him and Wallace. That also means Matthew conceived it before anyone else. Consequently, leading evolutionary scientists and biologists agree that only Matthew was the first with the entire theory (e.g. see: Wallace 1879a, de Beer 1962, Mayr 1982, Hamilton 2001, Cock and Forsdyke 2008, Wainwright 2008, Dawkins 2010, Rampino 2011, Ford 2011 and Weale 2015).

2. Darwin and Wallace replicated Matthew's original essential explanatory analogy of differences between artificial and natural selection, and more of his unique explanatory examples. Darwin four-word shuffled Matthew's original phrase. The words "process", "selection" and "natural" being essential to his theory of the "natural process of selection" and Darwin's "process of natural selection". The F2b2 hypothesis is supported by Darwin's four-word-shuffle, and is confirmed by Chambers (1859), who cited Matthew (Chambers 1832), he being apparently first to be second with Matthew's original phrase. Though proven fallible, the F2b2 hypothesis has, to date, been remarkably far more robust than originally expected (Sutton 2016).

3. Darwin, lied when he wrote that no naturalist (Darwin 1860a) and nobody at all (Darwin 1861a, 1861) had read Matthew's prior-published work. In fact, Matthew twice informed Darwin in published print in the Gardeners' Chronicle that the

exact opposite was true. Matthew (1860) explained to Darwin that the naturalist Loudon had reviewed and cited his book. Matthew (1860b) further explained that his work was heretical in the first half of the 19th century, so that even a professor of an esteemed university feared to teach it on penalty of the "cutty stool" (being pilloried on a three legged stool in church), and that Perth public library in Scotland banned his book for the same reason. Matthew informed Darwin also about the extensive United Services Journal (1831) review of his book. In that book review, the anonymous reviewer wrote *'...we disclaim participation in his ruminations on the law of nature.'* Despite the historical context of such brute censorship in the first half of the 19th century, Dawkins (2010) ignores such inconvenient facts and writes instead, without any apparent understanding of historical context of heresy in science, that Matthew never understood what he had originated or else he would have trumpeted it from the rooftops.

4. Matthew's (1831) original ideas on evolution by natural selection are in the main body of his book, as well as in its appendix. Darwin lied when he penned his "Hidden in an Appendix Fallacy" in every edition of the *Origin of Species* from the 1861 third edition onward. That lie has been regurgitated by countless Darwin scholars ever since. It is a proven lie, according to philosophical definition of what a lie is (Frankfurt 2005), because Matthew's (1860) first letter to the Gardeners' Chronicle on this issue included swathes of highly relevant natural selection text from the main body of his (1831) book. Darwin (1860e) wrote to his best friend, the famous economic botanist Joseph Hooker, that it would be "splitting hairs" to admit the truth of that particular matter.

5. Significant original research discoveries (Sutton 2014a) uncovered the independently verifiable fact that, as opposed the myth started by Darwin (1860a, 1861a, 1861) as the premise to support his and Wallace's claimed independent discoveries of Matthew's prior published theory, other naturalists in fact did read Matthew's original conception before he brought it to Darwin's attention in the Gardeners' Chronicle in 1860. This means we now newly know that knowledge contamination routes existed between those who

read Matthew's original ideas and the minds of Darwin and Wallace. There is no hard evidence that those routes were taken, but it could have occurred orally or in lost, destroyed or overlooked correspondence. For example, Darwin met and corresponded with Blyth, whose editor was Loudon, who cited *NTA*, and wrote that Matthew appeared to have something original to say on the "origin of species", no less. Darwin and Wallace knew William Hooker, who was a correspondent of Jameson, who cited *NTA*. Wallace's Sarawak paper editor was Selby, who cited *NTA*. Darwin's father was Selby's houseguest and Selby was the best friend of Darwin's great friend Jenyns and many others in his circle. In the 1840's, Darwin met and corresponded with Chambers, who cited *NTA* in 1832. Chambers then cited Matthew's (1839) second book *Emigration Fields* (Chambers 1840). Wallace was a correspondent of Darwin's mentor Lyell, who attended at least one meeting where Chambers gave a paper. In the 1840's, Both Darwin and Lyell wrote private letters revealing they thought they knew Chambers was the secret author of the *Vestiges*. Chambers (1844) was the anonymous author of *The Vestiges of Creation*. That book put evolution in the air in the first half of the 19th century, and paved the way for public acceptance of the heretical theory of evolution by natural selection. Both Darwin and Wallace admitted the *Vestiges* influenced society and their own work. Wallace wrote that the book was his greatest influence.

6. In fact, as opposed to Darwin's (1861a) lie that none at all read Matthew's original conception, at least 25 people cited Matthew's book in the literature many years before Darwin's and Wallace's (1858) papers were read before the Linnean Society. Seven naturalists read Matthew's original ideas pre-1858, because they are among the 25 who cited Matthew's book in the literature before that year. Darwin knew four of those naturalists.

7. Three of those four naturalists played major roles at the epicentre of influence on Darwin and Wallace and on their associates, influencers and influencer's influencers.

8. The three are John Claudius Loudon (1832), Prideaux John Selby (1842) and Robert Chambers (1832).

9. Loudon, who was best friends with the economic botanist John Lindley and William Hooker (father of Darwin's best friend Joseph Hooker), was owner and chief editor of the journal that published two of Blyth's (1835, 1836) most influential papers on organic evolution. Darwin fully admitted from the third edition of the *Origin of Species* (1861) onward that Blyth was his most useful and prolific informant on the topic.

10. Wallace (1879a) would later write that Matthew was one of the most original thinkers of the first half of the 19th century. Wallace should know the truth of that matter. Arguably, his recognition of Matthew's originality means, logically, that Wallace knew Matthew was one of the most original thinkers of the entire 19th century. That must make Matthew a reasonable candidate for recognition as one of the most original thinkers of all time.

~~~

Three main things allowed Darwin and Wallace to commit their plagiarizing science fraud by glory theft and get away with it. Firstly, they were facilitated by the 19th century norms of both the Royal Society and the British Association for the Advancement of Science. Those norms meant that political news, scientific arguments against the divine revelations of natural theology and deductive theories ought not be discussed or published (Gleick 2010). Their basis is in two Acts of Parliament of 1794 against sedition and treason (see Uglow 2002), which banned scientific societies from discussing scientific ideas that amounted to heresy, or included political issues. Matthew's (1831) book was completely interwoven with original ideas and scientific comments on such matters.

The resulting failure of all but a handful of naturalists to engage with Matthew in the literature of science in the first half of the 19th century made it easy for Darwin to claim that he was unaware of *NTA*'s existence, and to lie by writing that no other naturalist had read it either.

The scientific community sat back and did nothing other than permit Darwinists to decide upon the issue of whether or not the universally agreed originator of macroevolution by natural selection, incidentally not named Darwin, influenced anyone with his original published discovery. Having decided in their namesake's favour, Darwinists were then permitted to ignore the protocols of scientific priority by attributing Darwin with foremost priority for Matthew's prior-published breakthrough. They did so using a fallacious premise, started as a lie by Darwin, that no naturalist read Matthew's origination before 1860.

Matthew wove his discovery of natural selection throughout *NTA* to explain how best to grow suitable timber to supply Britain's burgeoning demands for shipbuilding, cities and chemicals essential for the industrial revolution. At the end of *NTA*, he concentrated further on his theory in an appendix where he included his heretical conclusion that, contrary to 19th century orthodox knowledge beliefs, a divine being had not designed and created species.

Matthew's big idea combined his Edinburgh University education, extensive European travel, 20 years of farming, fruit tree hybridizing and forestry experience, natural history observation, botanical and agricultural experimentaton. His work appears influenced by a distinctive understanding and improvement upon the radical evolutionary ideas of Buffon, Lamarck, Cuvier and de Candolle (see Dempster 2005). Those radical ideas came to a head in public debate in Paris in 1830, the year preceding publication of *NTA*.

In 1831, moral panic swept Britain in the wake of social unrest associated with the radically libertarian and reformist movement, of which Matthew was later to become an active leader as a Scottish regional representative of the Chartist movement. He used his understanding of natural selection to argue in *NTA* for social reform, and to mock the upper and middle classes. That clearly did him no favours at all with 19th century gentlemen of science.

Despite the fact that his big idea was religious heresy, Matthew went to the trouble of getting it right, no doubt writing draft after draft with an ink pen. Finally, he added more, by way of that dangerously condensed heretical summary of his conclusions in an appendix. If his book was to be banned for heresy or sedition, its publisher, bookseller or owner could at least remove the appendix in order to save the rest from the bonfire. Matthew then went to the immense personal risk of having *NTA* published in his name with not one, but two leading publishers: Black's of Edinburgh, and Longman and Co. of London. That fact alone made it anything but an obscure book.

Writing at a time when all British schools and universities taught that a worshipful supernatural being made the world in six days, along with all its rich diversity of life, Matthew had the intellect and personal courage to reject imagined truths explained by myths to fill the knowledge gaps about where the great variety of life comes from. In sum, he sought answers to the problem of species in natural science, rather than natural theology.

On April 7, 1860, the Gardeners' Chronicle published a lengthy letter from Matthew letting the world know that Charles Darwin had published his ideas without due attribution. On April 21, the Chronicle published Darwin's capitulation letter, which recognized Matthew's complete priority to the entire discovery and conceptual explanation of natural selection. In his letter, Darwin first apologized. Next, he set about excusing himself by claiming to have had no prior knowledge of Matthew's book or that the original ideas within it existed there. Then he further excused himself by claiming that, apparently, no other naturalist had heard of Matthew's original views, either. Darwin went on to add another three excuses. He claimed that Matthew's original views were only briefly given, that they appeared only in an appendix to his book and that the book was on naval timber and arboriculture, heavily implying that the subject was both obscure and of no obvious relevance to naturalists such as he researching the subject matter of organic evolution of species.

As we have seen, by reference to verifiable facts, the trouble with those accepted solutions to the historical problem of Darwin's claim to have independently discovered natural selection is that the very reasoning on which they are based is at odds with facts that were immediately discoverable at the time Darwin's reasoning was written. In short, had anyone bothered to question those defences and then examine the facts, each and every one of Darwin's excuses was capable of being refuted on the day he dishonestly fabricated them.

In that first Gardiners' Chronicle, published letter on his priority we know that Matthew (1860) wrote that the famous naturalist botanist, publisher and garden designer John Loudon had reviewed his book in print. Yet Darwin (1860a) replied that that he was not surprised that no naturalist had read Matthew's work. Why would he write such a thing, when Darwin knew Loudon was a famous naturalist? To be blunt, why would he tell such a whopping great lie?

Not much brainpower is required to see that plagiarizing science fraud by glory theft was Darwin's most likely motive.

Loudon, who died in 1843, was a member of the naturalist Linnean society. So was Darwin. In the first half of the 19th century, Loudon owned and edited 'The Magazine of Natural History', which frequently carried the strap line on its title pages that Loudon was a member of the Linnean society, the Geological Society, the Zoological Society of London and 'various other societies of the continent'. If Loudon was not a naturalist then nobody was.

Perhaps wishing simply to abide by the 19th century codes of courteous public conduct, Matthew ignored Darwin's falsehood about Loudon and instead politely exploited the opportunity that it presented to write of yet more naturalists who had read his original ideas. By way of his published reply to Darwin (1860a) on May 12, 1860, again in the Gardeners' Chronicle, Matthew (1860b) explained that his book had been banned for its heresy on the origin of species by the public lending library in Perth, Scotland (referring to it by its nickname the "Fair City"). He wrote forcefully that Darwin was wrong to assert that no naturalists had read his original ideas. He explained that an eminent university naturalist, who, out of obvious courtesy, he did not name, feared pillory punishment in the first half of the 19th century if ever he were to teach them to any students, or otherwise share Matthew's origination of the theory of evolution by natural selection.

Darwin's only response was to tell his lie again. He wrote the exact opposite of what Matthew informed him about other naturalists reading and understanding the significance of his original ideas. Pulling the curtain down on the stage of veracity, the more eminent, Darwin (Darwin 1861, 1861a) asserted next that no one whatsoever read Matthew's ideas. He conjured up the lie that they had gone unread before he and Wallace independently conceived them.

To consider the historic context ignored by cherry picking Darwin scholars, the title and subject matter of Matthew's prominently published and widely advertised book was far from obscure in the first half of the 19th century. Naval timber and arboriculture were respectively of top priority and high interest. For example, it took 5,000 oak trees to build a ship the size of the HMS Victory, and all such wooden ships of the age suffered constant ravages of dry rot and wreckage. Moreover, timber was the most important commodity of all for the industrial revolution, it being required for building, burning and, most essentially as a source of essential chemical components used in the manufacture of textiles. Timber, naval and otherwise, fuelled the industrial revolution. As for arboriculture, that was the passion of the landed gentry and city park designers. Matthew's book contained valuable lessons on the planting, re-planting and maintenance of beautiful specimen trees. It contained many lessons of great commercial and academic interest in the field of economic botany, a subject dominated by Darwin's best friend Joseph Hooker and his father William.

Far from being on a different and obscure topic to the interests of naturalists, the title and subject matter of *NTA* was, arguably, ideal for the inclusion of Matthew's theory and his scientific call for empirical research and scientific experimentation to test it.

Unlike Darwin, Matthew had no champions. Darwin inherited via his famous grandfather and father a direct route into being a fully networked, famous and influential member of the 19th century scientific community. He exuded an aura of respectability, honesty, diligence, focus and high intellect. His excuses to Matthew were taken completely at face value and without question. No one seemed to question their plausibility. Nevertheless, that should never have happened, because once that bridge was crossed, it followed logically that Matthew had not influenced Darwin. In addition, since Darwin had synthesized so much of the literature on the subject of species, and gathered a wealth of confirmatory examples from an amazing network of correspondents, it followed that the only reason he had not read or even heard of Matthew's book must have been that it was very obscure. In addition, so such reasoning goes, the book was surely obscure because such an important paradigm changing theory within it must have been obscurely located and obscurely written. How else could it be missed by both Darwin and Wallace?

Following such unrecognized fallacies to their natural conclusion, as a farce plays out, it was reasoned that Matthew had not only failed to influence Darwin but he had failed also to influence anyone else whatsoever who mattered. Moreover, from that it followed that the reason for those failures would be that it was his own fault, certainly not Darwin's, because it was not Wallace's fault either.

Darwinists are gatekeepers of knowledge about Darwin. This is due to their de facto "evolutionary expert" monopolization of the mainstream scientific literature on the history of the discovery of evolution by natural selection. To whom else, other than an expert Darwinist, would a rightfully cautious mainstream scientific literary agent, editor or publisher send a book proposal or article on the topic of Darwin and Matthew? This is a simple fact, not one to be confused with, or snidely accused of being, a "cover-up" conspiracy theory. To support that argument, I call upon what one celebrity scientist brilliantly sums up regarding what a conspiracy theory really is (Neil deGrasse Tyson 2011): *'Conspiracy theorists are those who claim coverups whenever insufficient data exists to support what they're sure is true.'*

Based on data in this book, the verifiable evidence suggests that, rather than it being a conspiracy of silence, most Darwin scholars are in a sociological state of denial. The sociologist Stanley Cohen (2001: p.1) explained for us what a state of denial is:

'One common thread runs through the many different stories of denial: people, organizations, governments or whole societies are presented with information that is too disturbing, threatening or anomalous to be fully absorbed or openly acknowledged. The information is therefore somehow repressed, disavowed, pushed aside or reinterpreted. Or else the information 'registers' well enough, but its implications - cognitive, emotional or moral - are evaded, neutralized or rationalized away.'

The international embarrassment of the Darwinists' 'state of denial' would have surely ended earlier had not control of the history of published academic thought on organic evolution been so completely entrusted to a group of devoted scientific Darwin cultists named after a proven lying post-hoc science fraudster by glory theft.

The verifiable evidence in this book is enough to suggest we should no longer reasonably expect such people to be capable of weighing objectively the verifiable evidence that their namesake committed science fraud by plagiarizing Matthew's theory and lying when he claimed no naturalist had prior knowledge of it.

The discoveries of most originators and first proposers do not come out of thin air. Rather, they represent some kind of problem solving breakthrough arising from an original, intellectual synthesis of existing knowledge, or else from new discoveries arising out of the observed outcome of experiments informed by old knowledge. The influence of Patrick Matthew's published discovery of natural selection before Darwin's *Origin,* is unlikely to be an exception given what we newly know about who read and then cited his work before

1858.

The public wants heroes they can understand and place within the scheme of things. In the words of Pardy (2012), they want to know that *"Newton discovered the laws of motion and Darwin discovered evolution."* What the public wants to know is one thing. Arguably, what the public actually needs to know, in a world dependent on veracity for greater success, is the truth, as opposed to the romance and lies.

The successful use of IDD to puncture the paradigm supporting the myth of Darwin and Wallace's dual immaculate conceptions of Matthew's prior published origination, confirms the pertinence of James Shapiro's advice on what would make the debate over Darwinism more productive (Shapiro1997, p.6):

The present debate over Darwinism will be more productive if it takes place in recognition of the fact that scientific advances are made not by canonizing our predecessors but by creating intellectual and technical opportunities for our successors.'

We may be about to experience a period of great intellectual uncertainty as new facts pour in to extinguish the flames held for other invented originations. All the better we celebrate true heroes than worship fraudsters and their profiteering acolytes.

Modern science requires proof to support a hypothesis. Darwin was motivated to spend decades gathering a wealth of evidence from the literature and his network of correspondents to prove Matthew's original breakthrough correct. Such an impressive number of confirmatory examples enabled him repeatedly, but wrongly, to refer to Matthew's prior-published discovery as "my theory," even after he admitted that it was Matthew's discovery.

Impressive, and indeed as crucial as all Darwin's wonderful collection of extra evidence is for confirmation of Matthew's theory, mere evidence gathering, no matter how prolific and important, can never magically transmute someone else's ideas into your own. Nevertheless, evidently, it can create just such a problematically long lasting mass illusion that it can.

Contrary to credulous unquestioningly dunderhead Darwinite rhetoric, it is arguable that Naval timber was brilliantly chosen, perhaps too brilliantly for other minds and mores of the time, as the perfect topic in which to situate the origination of the theory of macroevolution by natural selection. Page 1 of *NTA* supports this argument, with reference to how human varieties and our entire species invades the niches occupied by others Matthew (1831) wrote: *'NAVIGATION is of the first importance to the improvement and perfecting of the species in spreading, by emigration the superior varieties of man…'*

In 2014, the parting words of the first edition of this book (Sutton 2014a) were my final thoughts at the time on the story of Matthew, Wallace and Darwin. They belong to the botanist farmer who solved of the problem of species and influenced the world of science with his discovery (Matthew 1831, p.359):

'We have endeavoured to assist in disentangling the correct from the erroneous. It is impossible for the most wary always to avoid misconception of facts, but man merits the name of rational only, when he evinces a readiness to break from those misconceptions, to which the narrow minded, the proud, the vain, and the creature of habit and instinct, cling so obstinately.'

In 2017, I now follow that with a quotation from my first peer reviewed science article on the topic (Sutton 2015, p.204):

'Following up these leads, future research should focus on the archives of those naturalists who we now know — contrary to the previously unquestioned myths disseminated by leading scholars in the field — in fact did read Matthew's original ideas because they cited his book before 1858; similarly for their friends and correspondents. In the interests of a veracious history of scientific discovery, this proposed way forward for research would enable us to establish whether or not there is any discoverable record of mention being made, either to or from Darwin or Wallace, of Matthew and his original ideas on natural selection, before they each replicated them without citing him. Let us name the testable proposition, that such a note or letter will be found, the New Data-Led Hypothesis.'

7 REFERENCES, INCORPORATED WITH FULL BIBLIOGRAPHY OF REFERENCES FROM 1ST EDITION E-BOOK

Aaronovitch, A. (2010) Voodoo Histories: How Conspiracy Theory Shaped Modern History. London. Vintage.

A Society of Clergymen (1821) *The Washington theological repertory*. Volume 2. p. 275. Washington. J. Ashman.

Acta Germanica; or, The literary memoirs of Germany (1742). By a Society of Gentlemen. London. G. S. Smith

Agricola, G. A (1721) *A Philosophical Treatise of Husbandry and Gardening: Being a New Method of Cultivating and Increasing All Sorts of Trees, Shrubs, and Flowers*. London. P. Vaillant, and W. Mears and F. Clay.

Alexander, E. P. (1995) *Museum Masters: Their Museums and Their Influence*. London. Sage.

Alexander, R. D. (1974) The Evolution of Social Behaviour. Paper at the Institute on Religion in an Age of Science symposium "The Human Prospect: Heilbroner's Challenge to Religion and Science," Washington, D.C., October 23–24, 1974. Published in Johnston, R. F, Frank, P. W. and Michener, C. D. (eds.) *Annual review of ecology and systematics*. Volume 5. p.343.

Alexander, M. (2016) Perthshire Charles Darwin claims are 'so silly', claims leading international academic. *The Courier*. May 17th.

Allegedly Dawkins, R. (1993) Message from Dr. Richard Dawkins. Via Steven Brenner. Monday March 8th.http://www.bio.net/bionet/mm/mol-evol/1993-March/000781.html

Allen, D. E. (1967) J. F. M. Dovaston, an Overlooked Pioneer of Field Ornithology. *Journal of the Society for the Bibliography of Natural History*. Volume 4. pp.277-283.

Anonymous (1843) *Economic Uses of the Willow*.

Anonymous (1746) *The Museum*. Or the Literary and Historical Register. Vol. 2. London. R. Dodsley. Page 209.

Anonymous (1838) *The Penny Magazine* Economical uses of the Willow of the Society for the Diffusion of Useful Knowledge.

Anonymous (1812) *The Antijacobin Review and True Churchman's Magazine or Monthly, Political and Literary Censor*. From January to April Inclusive. With an Appendix Containing an Ample Review of Foreign Literature. Volume XLI. London. Craddock. page 327.

Anonymous (1826) Observations on the nature and Importance of Geology. *The Edinburgh NewPhilosophical Journal*. pp. 293-302.

Anonymous (1843) *The Penny Magazine of the Society for the Diffusion of Useful Knowledge*. London. Charles Knight and Co. Volume 11. pp. 434-436.

Anonymous (1844) Useful Applications of the Beech Tree. *The Penny Magazine of the Society for the Diffusion of Useful Knowledge*. London. Charles Knight and Co. Volume 12. pp. 159-160.

The Anthropological Review (1869) Volume 7. p. lxxxix

Arbesman, S. (2013) *The Half-Life of Facts: Why Everything We Know Has an Expiration Date*. New York. The Penguin Group.

Arcana of Science and Art (1832) Or an Annual Register of Popular Inventions and Improvements, Abridged from the *Transactions of Public Societies*, and from the *Scientific Journals, British and Foreign, of the Past Year*. Volume 5.

Arman, K. (2007) A new direction for kennel club regulations and breed standards. *Canadian Veterinary Journal*. September (48) 9 pp.953-965.

Andrews, C. (1853) *Reflections On The Operation Of The Present System Of Education*. Boston. Crosby Nichols and Company.

Anonymous (1701) *A rod for Tunbridge beaus, bundl'd up at the request of the Tunbridge Ladies*. To Jerk Fools into more Wit and Clowns onto more Manners. A Burlesque Poem. London.

Arnold, D. (2006) *The Tropics and the Travelling Gaze: India, landscape, and science*, 1800-1856. Seattle. University of Washington Press.

Arnott, J. W. (1848) *The London Gazette*.

The Athenaeum (1839) *Journal of Literature, Science and the Fine Arts*. January to December. p.17.

Auyang, S. Y. (1999) Foundations of Complex-system Theories: In *Economics, Evolutionary Biology, and Statistical Physics*. Cambridge University Press.

Avital, E. and Jablonka, E (2000) *Animal traditions: behavioural inheritance in evolution*, Cambridge, Cambridge University Press.

Aydinonat, N. E (2009) *How Economists Explain Unintended Social Consequences*. Taylor & Francis.

Asimov, I. (1983).*The roving mind*. Buffalo, N.Y: Prometheus Books.

Baddeley, F. (1829) Geology of a portion of the Labrador Coast. In *Transactions of the Literary and Historical Society of Quebec*. Vol. 1. Quebec/ Lemaitre.

Bailey, B. (2013) 'Bill Bailey's Jungle Hero'. BBC2. Television documentary in two parts.

Bajema, C. J., Drake, J. W. and Koch, R. E. (1976) *Eugenics: Then and now*. Stroudsburg, Pa. Dowden, Hutchinson and Ross.

Barck, H. (1762) *On the Foliation of Trees*.

Barbieri, M. (2003) *The organic codes: an introduction to semantic biology*. Cambridge. Cambridge University Press.

Barham, A (1831) *An address to earl Stanhope, on a survey of Chevening park made in 1817*. Self published.

Barnes, B., Bloor, D. and Henry, J. (1996) *Scientific Knowledge: A Sociological Analysis*. London. The Athlone Press.

Barrett, P. H., Gautrey, P. J., Herbert, S., Kohn, D., and Smith, S. (1987) Charles Darwin's Notebooks. Cambridge. Cambridge University Press.

Barruel, A. (1798) Memoirs, illustrating the history of Jacobinism: Part III *The Antisocial Conspiracy*. London . T. Burton.

Barton, R. (2013) 'X Club (act. 1864–1892)', *Oxford Dictionary of National Biography*, Oxford University Press, Oct 2006; online edition, May.

Bates, H. W. and Wallace, A. R. (1848) Manuscript letter to Sir William Hooker. 3rd April. Archives of the Royal Botanic Gardens. Kew.

Beccaloni, G. A (2008). Tour of Wallace's Places of Residence. In Smith, C. H. and Beccaloni, G. A. (eds) *Natural Selection and Beyond: The Intellectual Legacy of Alfred Russel Wallace*. Oxford. Oxford University Press.

Beddall, B. G. (1968) Wallace, Darwin and the Theory of Natural Selection: A Study in the Development of Ideas and Attitudes. *Journal of the History of Biology*. Vol.1. p. 261-324.

Beddall, B. G. (1972) Wallace, Darwin and Edward Blyth: Further notes on the development of evolution theory. *Journal of the History of Biology*. 5 (1) pp. 153-158.

Beer, G. Has Nature a Future? (1998) in Shaffer, E. S. (ed) *The Third Culture: Literature and Science*. Berlin. Walter de Gruyter and Co.

Bender, B. (2004) *Evolution And "the Sex Problem": American Narratives During The Eclipse of Darwinism*. Kent Ohio. Kent State University Press.

Bergman, J. (2011) *The Dark Side of Charles Darwin: a critical analysis of an icon of science*. Green Forest, Ark. Master Books.

Biagioli M (2012) From ciphers to confidentiality: secrecy, openness and priority in science. *British Journal for the History of Science* (45) Special issue (02): 213-33.

Bible (1611) Containing the Old Testament, and the New. Imprinted at London: By Robert Barker.

Bjorklund, D. F. and Blasi, C. H. (2011) *Child and Adolescent Development: An Integrated Approach*. Belmont. Wadsworth.

Black, R. (2003) Chimps genetically close to humans. *BBC Science News*. 20th May:

Blackmore, S. (2010) The Third Replicator. *New York Times*. Opinionator. August 22. p. 2.

Blair, G. (1845) *The Holocaust, or, The witch of Monzie: a poem, illustrative of the cruelties of superstition; Lays of Palestine, and other poems, to which is prefixed Enchantment disenchanted, or, A treatise on superstition*. London. J.F. Shaw.

Blume, L. and Durlauf, S. N. (2009) *Game theory*. Basingstoke: Palgrave Macmillan.

Blaine, D. (1817) *Canine Pathology*. Or a full description of the Diseases of Dogs. London. Boosey.

Blyth, E. 1835 An attempt to classify the "varieties" of animals. *The Magazine of Natural History*. (8) (1), Parts 1-2.

Blyth, E. (1836) Observations on the various seasonal and other external Changes which regularly take place in Birds more particularly in those which occur in Britain; with Remarks on their great Importance in indicating the true Affinities of Species; and upon the Natural System of Arrangement. *The Magazine of Natural History*: Volume 9. pp.393 – 409.

Boue, Dr (1834) On the Theory of the Elevation of Mountain Chains as advocated by M Elie de Beaumont. By Dr Boue. Communicated by the Author In Jameson, R. (1834) *The Edinburg New Philosophical Journal*. Vol 17. April-October. Edinburgh. Adam and Charles Black.

Boutcher, W. (1775) *A Treatise on Forest Trees*. London. Murray. J.

Bowler, P. J. (1983) *Evolution: the history of an idea*. Berkeley. The University of California Press. p.158.

Bowler, P. J. (2013) *Darwin Deleted: Imagining a World without Darwin*. Chicago. University of Chicago Press.

Bowler, P. (2007) Monkey Trials and Gorilla Sermons: Evolution and Christianity from Darwin to Intelligent Design. Harvard University Press.

Buffon, G. L. L. (1793) *The Natural History of Birds*. Volume 3. London. Murray.

Buffon, G. L. L. (1775) *The Natural History of Animals, Vegetables, and Minerals*. With the Theory of the Earth in General. Volume 3. London. T. Bell.

Burkhardt, F. H. (2001) Darwin and the Copley Medal. *Proceedings of the American Philosophical Society*. Vol. 145. No. 4. December. pp. 510-518.

Bradbury, A. (2005) Charles Darwin - The Truth? Website article: http://www.bradburyac.mistral.co.uk/dar0.html

Braun, D. P. (1990). Selection and evolution in non-hierarchical organization. In Upham, S. (ed) *The Evolution of Political Systems: Sociopolitics in Small- Scale*. Cambridge. Cambridge University Press.

Brewer's Dictionary of Phrase and Fable (2012) 19th Edition. London. Chambers Harap Publishers Ltd.

Brineley, T. (1993) *The Industrial Revolution and the Atlantic Economy: Selected Essays*. London. Routledge.

British Association for the Advancement of Science (1840) *Report of the Ninth Meeting of the British Association for the Advancement of Science*. Held at Birmingham in August 1839. London. John Murray.

British Association for the Advancement of Science (1868) *Report of the Thirty-Seventh Meeting of the British Association for the Advancement of Science*. Held at Dundee in September 1867. London. John Murray.

British Forest Trees (1843): A Guide to the Beauties of our Woods and parks. To which is Added the Natural History of Trees, with an Account of their Various Uses. London. Cradock and Co.

Brackman, A. C. (1980) *A delicate arrangement: the strange case of Charles Darwin and Alfred Russel Wallace*. New York, Times Books.

British Association for the Advancement of Science (1850) *Report of the Nineteenth Meeting held at Birmingham*. London. John Murray.

British Association for the Advancement of Science (1914). *Report of the Eighty-Fourth Meeting. Australia.* July 28-August 31. London. John Murray.

Britten, J. and Boulger, G. S. (1912) Some Little-Known British Botanists. *The Journal of Botany.* British and Foreign.

Brock, W. H. and Meadows, A. J. (1998) *The Lamp Of Learning: Taylor & Francis And Two Centuries Of Publishing.* London. Taylor and Francis. p. 123.

Brodie, J. (1867) *Papers offered for Discussion at The British Association for the Association for the Advancement of Science at Dundee,* In reply to the Speculations Recently Promulgated in Regard to the Antiquity and Nature of Man. Edinburgh: Johnson, Hinter and Co.

Broekhoff, J. (2005) Monkey Business: Evolution. Culture and Youth Sport. In Day, J. A. P. and Duguet, J. W. (eds.) *Kinanthropometry IV.* London. Taylor and Francis.

Brooks, J. L. (1999) *Just Before the Origin: Alfred Russel Wallace's Theory of Evolution.* Lincoln. Columbus University Press. pp. 258-268.

Brophy, A. L. (2009) Property and Progress: Antebellum Landscape Art and Property Law. *McGeorge Law Review.* Volume. 40. pp. 603-659.

Brown, A. (2009) *Haunted Kentucky: Ghosts and Strange Phenomena of the Bluegrass State.* Mechanicsburg. Stackpole Books.

Buchanan, R. (1859) *Notes of a Clerical Furlough Spent Chiefly in the Holy Land, with a sketch of the voyage out in the Yacht "St. Ursula"* London. Blackie and Son.

Buckland, W. (1836) The Bridgewater Treatises On the Power Wisdom and Goodness of God As Manifested in the Creation. Treatise VI. *Geology and Mineralogy Considered with Reference to Natural Theology.* In Two Volumes. Vol 1. London. William Pickering.

Buel, J. (1840) in Armstong, J. and Buel, J. (1840) *A Treatise on Agriculture.* New York. Haper and Brothers.

Burdge, R. (2017) Funding award will help to commemorate Carse man's scientific legacy. *The Courier.* May 9th.

Burrus, D. A. (2013) 3 Levels of Creativity: Leverage Your Unique Gifts. The Blog of Daniel A. Burrus. 11.12.2013. BestThinking.Com. http://www.bestthinking.com/thinkers/danielburrus?tab=blog&blogpostid =21568

Burt, A. and Trivers, R. (2009) *Genes in Conflict: The Biology of Selfish Genetic Elements.* Harvard University Press.

Butler, S. (1878) *Life and Habit.* London Trübner and Co.

Butler, S. (1886) *Luck, Or Cunning As the Main Means of Organic Modification.* Publisher Unknown.

Calman, W. T. (1912) Patrick Matthew (1790-1874) *The Journal of Botany.* British and Foreign. pp. 193-194.

Calman, W. T. (1912a) Patrick Matthew of Gourdiehill, Naturalist. *British Association, Dundee Meeting, 1912. Handbook.* David Winter and Son. Dundee. pp. 451-457.

Callioni, P. (2010) *Waves of change: managing global trends in the financial services industry.* Hawkhurst, Cranbook, Kent : Global Professional Pub.

Cambridge (1942) *Correspondence between members of the Darwin family,* British Association, etc., & University Library, Cambridge, on the gift of the Darwin MSS. (1942-49). DAR156. http://darwin-online.org.uk/content/frameset?itemID=CUL-DAR156&viewtype=image&pageseq=1

Campbell, B. G. (1998) *Human Evolution: An Introduction to Mans Adaptations.* New Jersey. Aldine Transaction, Rutgers. The State University.

Campbell, D. T. (1975). The Conflict between Social and Biological Evolution and the Concept of Original Sin. Zygon, *Journal of Religion and Science.* 10: 234–249.

Carey, T. V. (1998) *The Invisible Hand of Natural Selection and Vice Versa.* Springer.

Carleton, J. W. (1841) Review of: Selby, F. L. S. (1841) A History of British Forest Trees. Parts II and III. In: *The Sporting Review.* July. pp.376- 378.

Carleton, J. W. (1848) *The Natural History of the "Hawk" Tribe*. D. Bogue. London.

Carleton, J. W. [Craven] (1855) *Walker's manly exercises: containing rowing, sailing, riding, driving, racing, hunting, shooting and other manly sports*. London. H. G. Bohn.

Carpenter, H. (2008) *The Seven Lives of John Murray: The Story of a Publishing Dynasty 1768-2002*. London. John Murray.

Carroll, L. (1871) *Through the Looking-Glass, and What Alice Found There*. Macmillan.

Caton, H. (2007) Getting Our History Right: Six Errors about Darwin and His Influence. *Evolutionary Psychology*. Vol. 5 (1). pp.52-69.

Caven, B. (2014) Did Darwin copy ideas for Origin of Species? *The Scottish Daily Mail*. April 11. p.21.

Chalem, L. D. (2008) *Thrive with Diabetes*. Book Surge Publishing.

Chambers, R. (1832) *Lives of Illustrious and Distinguished Scotsmen from the Earliest to the Present Time Arranged In alphabetical order and forming a complete Scottish Biographical Dictionary* Part I (of eight volumes). Glasgow. Blackie and Son.

Chambers, W. and Chambers, R (1832). *Chambers's Edinburgh Journal*. William Orr. Saturday March 24th. p.63:

Chambers, R. (1832) In Chambers, W. and Chambers, R., *Chambers's Edinburgh Journal*. Vol. 1. Saturday November 3rd. pp.313-314.

Chambers, R. (1840) In Chambers, W. and Chambers, R. *Chambers's Edinburgh Journal*. Vol. III. page 40

Chambers, W. and Chambers, R. (1842). *Arboriculture*. In Chambers Information for the People. New and Improved Series. Part 8. Edinburgh. William and Robert Chambers. pp.401 to 416.

Chambers, R. (anonymous) (1844) *Vestiges of the Natural History of Creation*. New York. Wiley and Putnum.

Chambers, R. (1848) *Ancient Sea Margins as Memorials of Changes in the Relative Level of Sea and Land*. London. W and R Chambers.

Chambers, R. (1853) *Vestiges of the Natural History of Creation*. Tenth Edition. London. John Churchill.

Chambers, R. (1859) Charles Darwin on The Origin of Species. *Chambers's Journal of Popular Literature Science and Arts*. Saturday December 17. No. 311. pp.388-391. CUL-DAR121. Transcribed by Kees Rookmaaker. Darwin Online, http://darwin-online.org.uk/

Cock, A. G. and Forsdyke, D. R. (2008) Treasure Your Exceptions: The Science and Life of William Bateson. Springer.

Cohen, S. (2001) *States of Denial: Knowing About Atrocities and Suffering*. Cambridge. Polity Press.

Colp, R. (1977) To be an invalid: the illness of Charles Darwin. Chicago. University of Chicago Press.

Corbaux, F. (1829) On the laws of Mortality and the Intensity of Human Life. The Philosophical Magazine. Vol. 5. Pp. 198-205.

Costa, J. T. (2014) *Wallace, Darwin and the Origin of Species*. Cambridge Massachusetts. Harvard University Press.

Darwin, C. R. (1838) Books Read and Books to Read Notebook. Darwin Online:
http://darwinonline.org.uk/content/frameset?keywords=dar119%20cul&pageseq=1&itemID=CUL-DAR119.-&viewtype=text

Darwin, C. R. (1839) Observations of the Parallel Roads of Glen Roy and Other Parts of Lochaber in Scotland with an Attempt to Prove that they are of Marine Origin. *Philosophical Transactions of the Royal Society of London*. Volume 129. London. The Society.

Darwin, C. (1839a) *Voyages of the Adventure and Beagle*. Volume III. Journal and Remarks 1832-1836. London. Henry Colburn.

Darwin. C. R. (1843) Letter to John Lindley. 8th April. Darwin Correspondence Project. Letter no. 668.

Darwin, C. R. (1841a) Index to annotations by Darwin in his copies of Gardeners' Chronicle. Darwin Online Archive, Cambridge University Library. DAR 222.

Darwin, C. R. (1844) Letter to Joseph Hooker, J. D. 11 January. Darwin Correspondence Database.

Darwin, C. R. (1844a) Letter to Hooker, J. D. 22 July. Darwin Correspondence Database.

Darwin, C. R. (1844) Letter to Hooker. 10th November. Darwin Correspondence Database.

Darwin, C. R. to Hooker, J. D. (1845) Letter of 29 Aug. Darwin Correspondence Database.

Darwin, C. R. (1845b) Letter to Hooker. 8th October. Darwin Correspondence Database. Darwin Correspondence Database,

Darwin, C. (1845c). *Journal of Researches into the Natural History and Geology of the Countries Visited during the Voyage of H.M.S. Beagle round the World*. 2nd ed. London: John Murray.

Darwin, C. R. (1847) Letter to Hooker, J. D. 18 April. Darwin Correspondence Database.

Darwin, C. R. (1847a) Letter to Joseph Hooker. 5th October. Darwin Correspondence Database

Darwin, C. R. (1847b) Letter to Robert Chambers. 28th February. Darwin Correspondence Project. Letter no. 1065.

Darwin, C. R. (1848) Letter to Robert Chambers. 14th February-20th March. Darwin Correspondence Project letter no. 1160.

Darwin CR (1848a) Letter from Darwin, C. R. to Joseph Hooker. 10 May. Darwin Correspondence Project.

Darwin, C. R. (1849) 29th January. Letter to Strickland, H. E. Darwin Correspondence Database:

Darwin, C. R. (1849a) Letter to Strickland, H. E. 4th February. Darwin Correspondence Database.

Darwin, C. R. (1849b) Letter from Darwin, C. R. to Hooker, J. D. on 9 Apr 1849. Cambridge Digital Library:

Darwin, C. R. (1857) Letter to Asa Gray, September 5th, 1857. See Darwin online.org and for a clearer picture: The Unofficial Stephen J Gould Archive:

Darwin, C. R. (1857a) Letter to Sharpy. 9th April 1857. Darwin Correspondence Database.

Darwin, C. R. (1857b) Letter to Alfred Wallace. 1 May. See Darwin Correspondence Project. Letter number 2086.

Darwin, C. R. (1858) Letter to Lyell. 18th June 1858. See Darwin Correspondence Database.

Darwin, C. R. and Wallace, A. R. (1858) On the tendency of species to form varieties; and on the perpetuation of varieties and species by natural means of selection. *Journal of the Proceedings of the Linnean Society of London.*

Darwin. C. R. (1859) *On the Origin of Species by Means of Natural Selection.* Or the Preservation of Favoured Races in the Struggle for Life. London. John Murray.

Darwin, C. R. (1859a) Letter to Lyell. 30th March: Darwin Correspondence Database.

Darwin, C. R. (1859b) Letter to Hooker, J. D. 25 Dec Darwin Correspondence Database.

Darwin, C. R. (1859c) Letter to Wallace. 25th January. Darwin Correspondence Database.

Darwin, C. R. (1859d) Letter to Hooker 25th January. Darwin Correspondence Project. Darwin Correspondence Database

Darwin, C. R. (1860). In de Beer, G. ed. (1960). Darwin's notebooks on transmutation of species. Part I. First notebook [B] (July 1837-February 1838). *Bulletin of the British Museum (Natural History).* Historical Series 2, No. 2 (January): 23-73. Darwin Online.

Darwin, C. R. (1860a) Natural selection. *Gardeners' Chronicle and Agricultural Gazette* no.16 (21 April): pp.362-363.

Darwin, C. R. (1860b) Darwin, C. R. to Lyell, Charles. 10 Apr [1860] Letter. See Darwin Correspondence Project.

Darwin, C. R. (1860c) Letter to Baden Powell 18th January . Darwin Correspondence Project:

Darwin, C. R (1860d) Additional Note to Letter to Baden Powell of 18th January. Darwin Correspondence Project:

Darwin, C. (1860e) Letter to Hooker. 13th April. Darwin Correspondence Project. Darwin Correspondence Database,

Darwin, C. (1860f) Letter to Gray. 25th April. Darwin Correspondence Project.

Darwin, C. (1860g) Letter to Wallace, 18th May. Darwin Correspondence Project.

Darwin, C. (1860h) Letter to Hooker, 29th December. Darwin Correspondence Project.

Darwin, C. R. (1861) *On the Origin of Species by Means of Natural Selection.* Or the Preservation of Favoured Races in the Struggle for Life. (Third Edition) London. John Murray.

Darwin, C. R. (1861a) Letter to Qatrefages de Bréau, J. L. A. de. Apr. Darwin Correspondence Project.

Darwin, C. R. (1861b) Letter to Gray. 5th June. Darwin Correspondence Project:

Darwin, C. R. (1862) Letter to Asa Gray. 16th October, Darwin Correspondence Database.

Darwin, E. (1863) Letter to Patrick Matthew. 21 November. Darwin Correspondence Project.

Darwin, C. R. (1865) Letter to Hooker dated 22 and 28th October. Darwin Correspondence Project.

Darwin, C. R. (1865a) Letter to Huxley dated 12 July. Darwin Correspondence Project.

Darwin. C. R. (1865b) Letter to Huxley dated 17th July. Darwin Correspondence Project.

Darwin, C. R. (1868) The Variation of Animals and Plants Under Domestication. Volume 2. London. John Murray.

Darwin, C. R. (1868) *The Variation of Animals and Plants under Domestication.* London: John Murray. 2nd edition. Volume 1.

Darwin, C. R. (1870) Letter to Asa Gray. Dated 15th March. Darwin Correspondence Project.

Darwin, C. R. (1872) *On the origin of Species by Means of Natural Selection* (6th edition with addition of Historical Sketch). London. John Murray.

Darwin, Francis (1887) (ed). *The life and letters of Charles Darwin, including an autobiographical chapter.* London: John Murray. Volume 1.

Darwin, Francis (1887) (ed). *The life and letters of Charles Darwin, including an autobiographical chapter.* London: John Murray. Volume 2.

Darwin, C. R. (1896) *The life and letters of Charles Darwin: including an autobiographical chapter.* Edited by Darwin, F. New York. D. Appleton.

Darwin, C. R. (1909) *The Foundations of the Origin of Species*: Two Essays Written on 1842 and 1844. (Edited by Darwin, F.). Cambridge. Cambridge University Press.

Darwin, C. R. (1991) *The Correspondence of Charles Darwin*: Volume 7, 1858 1859. Cambridge, Cambridge University Press.

Darwin, C. and Glick, T. F. (1996) *On Evolution: the development of the theory of natural selection.* Indianapolis, Ind. Hackett.

Darwin, C. R. and Costa, J. T. (2009) The Annotated Origin: a facsimile of the first edition of *On the origin of species.* Cambridge, Mass: Belknap Press of Harvard University Press.

Dasgupta, S. (1994) *Creativity in Invention and Design: Computational and Cognitive Explorations of Technological Originality.* Cambridge. Cambridge University Press.

Davies, R. (2008) *The Darwin Conspiracy: Origins of a Scientific Crime*. London. Golden Square Books.

Dawkins, R. (1976) *The Selfish Gene*. (first edition) Oxford. Oxford University Press.

Dawkins, R (1979) Twelve Misunderstandings of Kin Selection. *Z. Tierpsychol.*, 51, 184—200.

Dawkins, R (2003) *A Devils Chaplain: selected essays*. London. Weidenfeld and Nicolson.

Dawkins, R. (2006) *The Selfish Gene*. 30th Anniversary Edition. Oxford. Oxford University Press.

Dawkins, R. (2006) *The God Delusion*. London. Bantam Press.

Dawkins, R. (2004) *The Ancestor's Tale: A Pilgrimage to the Dawn of Life*. London. Orion Books.

Dawkins, R. (2008) Why Darwin Matters. *The Guardian*. 9th February.

Dawkins, R. (2010) Darwin's Five Bridges: The Way to Natural Selection. In Bryson, B (ed.) *Seeing Further: The Story of Science and the Royal Society*. pp. 203-228. London. Harper Collins.

de Beer G (1962) *The Wilkins Lecture*: The Origins of Darwin's Ideas on Evolution and Natural Selection. Proceedings of the Royal Society of London. Series B. Biological Sciences 155 (960). pp.321-338.

de Humboldt, A. (1822) *Political Essay on the Kingdom of New Spain*. Vol. 3. London. Longman, Hurst, Rees, Orme and Brown.

deGrasse Tyson (2011), @neilty Twitter. April 7th. https://twitter.com/neiltyson/status/56010861382336513?lang=en-gb.

de Grey, A. (2009) Postponing Ageing: Re-Identifying the Experts in Healey, P. and Rayner, S (eds) *Unnatural Selection: The Challenges of Engineering Tomorrow's People*. Abingdon. Earthscan.

de Maillet, Benoît (1968) Teliamed, or conversations between an Indian philosopher and a French missionary on the diminution of the sea. Translated and edited by Albert V. Carozzi. University of Illinois Press, Urbana, Chicago & London.

de Saint-Pierre, J. H. B. (1796) *Studies of Nature*. Volume III. London. C. Dilly in the Poultry.

Deacon, T. W. (2011) *Incomplete Nature: How Mind Emerged From Matter*. W. W. Norton and Company.

Deely, J. N. (1969) *The philosophical dimensions of the Origin of Species*. Chicago. Institute for Philosophical Research.

Dempster, W. J. (1983) *Patrick Matthew and Natural Selection*. Edinburgh. Paul Harris Publishing.

Dempster, W. J. (1996) *Evolutionary Concepts in the Nineteenth Century*. Edinburgh. The Pentland Press.

Dempster, W. J. (2005) *The Illustrious Hunter and the Darwins*. Sussex. Book Guild Publishing.

Dennett, D. C. (1996) *Darwin's Dangerous Idea: Evolution and the Meanings of Life*. Great Britain. The Penguin Press.

Dercole, F. and Rinaldi, S. (2008) *Analysis of Evolutionary Processes: The Adaptive Dynamics Approach and Its Applications*. Princeton New Jersey. Princeton University Press.

Desmond, A. (1989) Lamarckism and democracy: Corporations, corruption and comparative anatomy in the 1830s. In Moore, J. R. (ed.) *History, Humanity and Evolution: Essays for John C. Green*. Cambridge. Cambridge University Press.

Desmond, A. (1989a) *The Politics of Evolution: Morphology, Medicine, and Reform in Radical London*. Chicago. University of Chicago Press.

Desmond, A. Moore, J. and Browne, J. (2007) *Charles Darwin*. Oxford. Oxford University Press.

Deutsch, D. (2011) *The Beginning of Infinity: Explanations that Transform the World*. London. Allen Lane: Penguin Books.

Dickens, C. (1860) *All the Year Round: A Weekly Journal*. Conducted by Charles Dickens. Volume III. April 14 to October 6 1860. No.s.51 to 76. pp.174-178. See also pp.293 to 299.

Discover Magazine (2006) 25 Greatest Science Books of All Time. Discover presents the essential reading list for anyone interested in science. By the editors of DISCOVER magazine. Friday, December 08. http://discovermagazine.com/2006/dec/25-greatest-science-books#.UfkaodLVCSo

Dobriansky, L. E. (1957) *Veblenism: a new critique*. Washington. Public Affairs Press.

Donaldson, J. (1796) *Modern Agriculture, Or, The Present State of Husbandry in Great Britain: Including an Account of the Best Modes of Cultivation Practised Throughout the Island, the Obstacles to Further Improvements, and the Means by which These May be Most Effectually Removed*, Volume 3. Edinburgh. Adam Neil and Company.

Dovaston, J. F. (1837) (writing under pseudonym von Osdat). Some Observations on the Oak. *Magazine of Natural History and Journal of Zoology, Botany, Mineralogy, Geology, and Meteorology*. Vol. 1. pp.74-77.

Dowd, W. (2013) A Room with a Conspiratorial View. *eSkeptic Magazine*. June 19th 2013: http://www.skeptic.com/eskeptic/13-06-19/

Dower, H. (2009) Darwin's Guilty Secret. Hughdower.com http://www.hughdower.com/guilty.html

Drayton, R. H. (2000) *Nature's Government: Science, Imperial Britain, and the 'Improvement' of the World*. Yale University Press.

Drayton, R (2009) 'Lindley, John (1799–1865)', *Oxford Dictionary of National Biography*, Oxford University Press, 2004; online edition, May http://www.oxforddnb.com/view/article/16674 Accessed 27 July 2013.

Drake, D. (ed.) (1834) *The Western Journal of the Medical & Physical Sciences*. Vol.7. Cincinnati Ohio. E. Deming.

Drowne, K. M. and Huber, P. (2004) *The 1920s*. Westport, Conn. Greenwood Press.

The Dublin University Magazine (1860). Vol. 55. January to June. pp.717-718.

Edinburgh Gazetteer (1822) *The Edinburgh Gazetteer* Or, Geographical Dictionary: Containing a Description of the Various Countries, Kingdoms, States, Cities, Towns, Mountains, &c. of the World; an Account of the Government, Customs, and Religion of the Inhabitants; the Boundaries and Natural Productions of Each Country, &c. &c. Forming a Complete Body of Geography, Physical, Political, Statistical, and Commercial.

The Edinburgh Literary Journal (1830) or, Weekly register of criticism and belles-lettres. Volume 4. Saturday December 18th. p.49. (Advertisement for the book *NTA*).

The Edinburgh Literary Journal (1831) or, Weekly register of criticism and belles-lettres. Literarily Criticism of Naval Timber and Arboriculture. July 2nd N. 138. pp.1-4.

Edwards, R. (1991) Cairngorm. Article Filed for *The Guardian*, 27 March. http://www.robedwards.com/files/CAIRNGORM.doc

Edwards, G. (1751) *A Natural History of Uncommon Birds*. Part IV and Last. London. Self published.

Eiseley, L. (1959) Darwin's Century. *Evolution and the Men who Discovered it*. London. The Scientific Book Guild.

Eiseley, L. (1979) Darwin and the Mysterious Mr X: *New Light on the Evolutionists*. New York. E. P. Dutton.

Eiseley, L. C. and Grote, A. (1959) Charles Darwin, Edward Blyth, and the Theory of Natural Selection. *Proceedings of the American Philosophical Society* Vol. 103, No. 1 (Feb. 28,) pp.94-158.

Eisenberg, D. and Campbell, B. (2011) The Evolution of ADHD, Social Context Matters. *San Francisco Medicine*. October. pp.21-22.

Eliot, A., Hood, T. and Broderip, F. (1885) *Hood in Scotland: reminiscences of Thomas Hood, poet and humorist*, Dundee. J.P. Mathew and co.

Elliott, B. (2010) The Reception of Charles Darwin in the British Horticultural Press. Occasional *Papers from the RHS Lindley Library*. 3. pp.5-83.

Emmons. E. (1846) *American Journal of Agriculture and Science*, Volumes 3-4. January to June p. 198.

Endersby, J. (2008) Joseph Hooker: a philosophical botanist. *Journal of Biosci.* Vol. 33. pp. 163-169.

Engels, E. (2008) The Reception of Charles Darwin in Europe: *The Reception of British Authors in Europe*, Volume 1. London: Continuum.

Epstein, J. (1982) *The Lion of Freedom: Feargus O'Connor and the Chartist Movement, 1832-1842.* p. 143. Bekenham. Croom-Helm.

Evelyn, J. (1664) *Sylva, or a discourse of forest-trees, and the propagation of timber.* To which is annexed Pomona; or an appendix concerning fruit trees in relation to cider. London. Jo. Martyn, and Ja. Allestry, printers to the Royal Society.

Faggen, R. (1985, p.277) *The Fact is the Sweetest Dream. Darwin Pragmatism and Poetic Justice.* In Bloom, H. (ed) (1985, p.294) Robert Frost: Revolution in science. Cambridge, Mass. Harvard University Press.

Fan, F. (2009) *British naturalists in Qing China: science, empire, and cultural encounter.* Cambridge, Mass. Harvard University Press.

Feeley-Harnik, G. (2004) The Geography of Descent, *Proceedings of the British Academy* Volume 125. 2003 Lectures. Edited by the British Academy. Oxford. Oxford University Press. p.322.

Feistel, R. and Ebeling, W. (2011) *Physics of Self-Organization and Evolution.* Weinheim. John Riley and Sons.

Feynman, R. P. (1992) "Surely You're Joking, Mr. Feynman!":*Adventures of a Curious Character.* London. Random House.

Fisher, G. P., Adams, G. B. and Farnam, H. W. (1909) *The Yale Review.* Vol. 17. p.131.

Fishbourne, E. G. (1855) *Impressions of China and the present Revolution its Progress and Prospects.* London. Seeley and Co.

Fitzroy, R. (1839) Voyages of the Adventure and Beagle. Volume II. *Proceedings of the Second Expedition.* London. Henry Colburn.

Flint, K. (1995) Origins, Species and Great Expectations. In Amigoni, D. and Wallace, J. (eds) *Charles Darwin's the Origin of Species: New Interdisciplinary Essays.* Manchester. Manchester University Press.

Fletcher, C. 1984. Why one man became the world hero. Review of Macfarlane, G. (1984) Alexander Fleming: the Man and Myth. *New Scientist.* March 22. p.30.

Floy, J. (1858) *The National Magazine: Devoted to Literature, Art, and Religion,* Volumes 12-13. p.183.

Fodor, J. and Piattelli-Palmarini, M. (2011) *What Darwin Got Wrong.* Exmouth. Profile Books.

Folwell, W. (1930) *A History of Minnesota.* Vol. IV. Minnesota Historical Society Press.

Ford, B. J. (2011) Darwin: The Microscopist Who Didn't Discover Evolution. *The Microscope.* 59: 3, pp.129-137.

Foster, J. B. (2000) *Marx's Ecology, Materialism and Nature.* New York. Monthly Review Press.

Fortey, R. (2010) Archives of Life: Science and Collections: In Bryson, B. (ed.) Seeing Further: *The Story of Science and the Royal Society.* London Harper Collins.

Forsyth, W. (1791) *Observations on the Diseases, Defects, and Injuries in All Kinds of Fruit and Forest Trees.* London.

Forsyth, M. (2011) *The Etymologicon.* London. Icon Books.

Fox, W. J. (1845) *Lectures, addressed chiefly to the working classes.* Volume 2. Page 22.

Francis, K. (2007) *Charles Darwin and The Origin of Species.* Westport CT. Greenwood Publishing.

Frankfurt, H. G. (2005) *On Bullshit.* Princeton. Princeton University Press.

Friedman, G. M. (1998): *Lyell, C in New York State.* Geological Society, London, Special Publications. Vol.143. pp.71-81

The Friend: A Religious and Literary Journal (1842) Vol. 15. p.308.

Fraser's Magazine for Town and Country (1841) Volume 23, June. Issues 133-138. Page 130.

Freeman, R. B. (2007) *Charles Darwin: A companion*. 2d online ed. compiled by Sue Asscher (edited by John van Wyhe).

Froude, R. H. (1838) *Remains of the late Reverend Richard Hurrell Froude*. London. J.G. and F. Rivington.

The Gardener's Magazine (1832) Vol. 8. London. Longman, Rees, Orme, Brown, Green and Longman.

The Garden (1884): An Illustrated Weekly Journal of Gardening in All Its Branches, Volume 25.

Gardner, D. (2010) *Future Babble*. London. Virgin Books.

Gauger, W. H. (1687) *The Gaugers Magazine*. London. Mary Clark.

Gayon, J (1998, p. 50) *Darwinism's struggle for survival: heredity and the hypothesis of natural selection*. Cambridge. Cambridge University Press.

Gazlay, A. (Pseudonym Broadluck, C). (1856) *Races of Mankind: With Travels in Grubland*. Cincinnati. Langley Brothers.

Geikie, A. (1912) Charles Lyell and Forfarshire Geology. British Association, Dundee Meeting, 1912. Handbook. David Winter and Son. Dundee. Ppp.416-422.

The Gentleman's Magazine (1834) vol. 1. January-June. p.459.

Ghiselin, M. T. (2003) *The triumph of the Darwinian method*. Mineola N.Y: Dover.

George, C. R. P. (1996) William Charles Wells (1757-1815)—a nephrologist of the Scottish enlightenment. *Nephrol Dial Transplant*. Historical Note. Vol 11. pp.2513-2517.

German, J. (1964) DNA Synthesis in Human Blood Cell Chromosomes. *The Journal of Cell Biology*. Vol. 20. pp.37 -55.

Gibson, J. P. and Gibson, T. R. (2009) *Natural Selection*. New York. Infobase Publishing.

Gilovitch, T. (2008) How We Know What Isn't So: The Fallibility of Human Reason in Everyday Life. New York. Simon and Schuster.

Gontier, N. (2010) Evolutionary epistemology as a scientific method: a new look upon the units and levels of evolution debate. *Theory Biosci.* 129: pp.167–182.

Gleick, J. (2010) At the Beginning: More things in Heaven and Earth. In Bryson, B. (ed.) *Seeing Further: The Story of Science and the Royal Society*. London. Harper Collins.

Goodman, D. (1985) Buffon's *Histoire naturelle* as a Work of the Enlightenment. In North, J. D. and Roche, J. J. (eds) *The Light of Nature: Essays in the History and Philosophy of Science*. Dordrecht. Martinus Nijhoff.

Gorrie, A. (1832) Account of the Carse of Gowrie. *Prize-Essays and Transactions of the Highland Society of Scotland.* Vol 9. Edinburgh. William Blackwood.

Gould, S. J. and Eldredge, N. (1977). 'Punctuated equilibria: the tempo and mode of evolution reconsidered.' *Paleobiology* 3(2): 115-151. (p.145).

Gould, S. J. (1983) Unorthodoxies in the First Formulation of Natural Selection. *Evolution.* Vol. 37, No. 4 July. pp.856-858.

Gould, S. J. (2002) *The Structure of Evolutionary Theory*. Harvard. Harvard University Press. pp. 137-141.

Grafen, A. in Grafen, A. and Ridely, M. (eds) (2006) *Richard Dawkins: How a scientist changed the way we think*. Oxford. Oxford University Press.

Grande, L. (2003) *The Lost World of Fossil Lake: Snapshots from Deep Time.* Chicago. University of Chicago Press. See page 18.

Grandin, T. and Deesing, M. J. (2013) *Genetics and the Behavior of Domestic Animals*. Academic Press.

Gräslund, B. (2005) *Early Humans And Their World*. Abingdon. Routledge.

Gray, S.(1815) *The Happiness of States*: Or an Inquiry Concerning Population, the Modes of Subsisting and Employing it, and the Effects of all on Human Happiness. London. Hatchard.

Grant, A. (1897) Spencer and Darwin. *Appleton's Popular Science Monthly*. April. pp.815-827.

Gray, A. (1859) *The American Journal of Science and Arts*. Second Series. Vol. 77. p.442.

Green, S. and Ray, D. (2009) *Potential impacts of drought and disease on forestry in Scotland. Research note*. Forestry Commission. Forest Research. Northern Research Station. Roslin. Midlothian. Scotland. September.

Gross, R. (2013) *Being Human: Psychological and Philosophical Perspectives*. Abingdon. Routledge.

Grothe, A. (1878) *The Tay Bridge: Its History and Construction*. Dundee, J. Leng and Co.

Gruber, H. E. and Barrett, P. H. (1974) *Darwin on man: a psychological study of scientific creativity*. London. Wildwood House.

Hale, M. H. (1869) *Sunshine and shadow in New York*. Hartford. J. B. Burr and company.

Hale, M. H. (1871) *Twenty years among the bulls and bears of Wall street*. Hartford. J. B. Burr and company.

Hall, J. (1807) *Travels in Scotland by an Unusual Route with a Trip to the Orkneys and Hebrides*, Containing Hints for Improvements in Agriculture and Commerce. London. J. Johnson. Vol 1.

Hall, A. R. (1996) *Henry More and the Scientific Revolution*. Cambridge. Cambridge University Press.

Hallier, E. (1866) *Botanische Zeitung*. 7th December. Volume 24. No.49. pp.382-383.

Hamilton, W. D. (1971) Selection of Selfish and Altruistic Behaviour in Some Extreme Models. Paper delivered at the Smithsonian Institution Annual Symposium 14–16 May 1969. In Eisenberg, J. F., Dillon, W. S. (eds) *Smithsonian Annual III*. Man and Beast: Comparative Social Behaviour. Washington. Smithsonian Institution Press.

Hamilton, W. D. (2001) Narrow Roads of Gene Land, Volume 2: *Evolution of Sex*. Oxford. Oxford University Press.

Hansen and Curtis (2010) *Voyages in World History*. Wadsworth. Cengage Learning.

Hargreaves Heap, S. P. and Varoufakis, Y. (2004) *Game Theory A Critical Introduction*, 2nd Edition. London. Routledge.

Harris, S. N. (1848) Intermittent and Remittent Fever. *Charleston Medical Journal and Review*. Vol. 3.

Harvati, K. (2012) What Happened to the Neanderthals? *Nature Education Knowledge*. 3 (10): 13.

Harvard University (2013) Jameson William (1796-1873) *Papers of William Jameson, 1827-1869*: A Guide. Archives, Gray Herbarium Library. Harvard University, Cambridge, MA. USA.

Harway, M. and O'Neil, J. M. (1999) *What causes men's violence against women?* Thousand Oaks, Calif: Sage Publications.

Hasan, H. (2005). *Mendel and the Laws of Genetics*. New York. The Rosen Publishing Group.

Hayden, T. (2009) What Darwin Didn't Know. *Smithsonian Magazine*, February.

Hayes, C. W. (2007) *Historic Orchards of the Carse of Gowrie*. Phase 1 Survey. An Investigative Study, On their Location, Extent and Condition. Report to Perth and Kinross Countryside Trust.

Hayes, C. W. (2008) Ancient Orchards on the Banks of the River Tay. *Landscape Archaeology and Ecology*, Volume 7. July. pp.63-75.

Head, G. (1829) *Forest scenes and incidents, in the wilds of North America: being a diary of a winter's route from Halifax to the Canadas, and during four months residence in the woods on the borders of Lakes Huron and Simcoe.* London, J. Murray.

Heister, L. (1750) *A General System of Surgery in Three Parts.* London. W. Innys.

Hesketh, I. (2009) *Of Apes and Ancestors: Evolution, Christianity, and the Oxford Debate.* Toronto. University of Toronto Press.

Heywood, J. (ed) (1840) *Collection of statutes for the University and colleges of Cambridge: including various early documents.* London. William Clowes and Sons.

Hildebrand, A. F. (1903) *A Voice from the Wilderness: Meditations of a Google.* San Francisco, California.

Hillhouse, A. L. (1818) *Description of the European olive tree.* Paris.

Hill, I. (ed) (1841) Visit to a Massachusetts Farm. *The Farmer's Monthly Visitor.* Volume 3. pp.136-139 (see page 138).

Hill's Raleigh (1951) *Wake County, N.C. City Directory.* Volume 39. Hill Directory Company, p.426.

Hindle, E. (1958) Darwin's Greatest Work. *The New Scientist.* 26th June. Vol. 4. No. 84. pp. 246-248.

Hinton, J. (1809) Causes of the Overthrow of the Spanish Monarchy By the Rev JOSETH TOWNSEND A Author of Travels in Spain. *The Universal Magazine of Knowledge and Pleasure.*

Hocken, T. M. and Johnstone, A. H. (1909) *A bibliography of the literature relating to New Zealand.* Wellington, N.Z., Printed by J. Mackay, government printer.

Hodgson, G, M. and Knudsen, T. (2010) *Darwin's conjecture: the search for general principles of social and economic evolution* Chicago; London : University of Chicago Press.

Hogg, R. (1859) *The apple and its varieties being a history and description of the varieties of apples cultivated in the gardens and orchards of Great Britain.* London. Groombridge and sons.

Holan, S. R., Gardos, G. and Sarkadi, B. (1980) *proceedings of the 28th International Congress of Physiological Sciences*, Budapest, 1980. Elmsford, N.Y.: Pergamon; Budapest: Akade☐ miai Kiado☐ .

Holmes, R. (2010) A New Age of Flight: Joseph Banks Goes Ballooning. In Bryson, B. (ed.) *Seeing Further: The Story of Science and the Royal Society.* London. Harper Collins.

Hook, E. B. (2002) *Prematurity in scientific discovery: on resistance and neglect.* Berkeley, Calif: University of California Press.

Hooker, W. J. and Greville, R. K. (1831) *Icones filicum etc. Figures and descriptions of ferns etc.* In Two Volumes. London.

Hooker, J. (1841) Works Written Or Edited by J.C. Loudon and Published On His Own Account, *The Gardeners' Chronicle* 1841, vol.1, no.44, p.714.

Hooker, J. (1843) Letter to Darwin. 28th November. Darwin Correspondence Project.

Hooker, J. D. (1844) Letter to Darwin. 28th October. Darwin Correspondence Database

Hooker, J. D. (1845) Letter to Darwin, C. R. 14 Sept. Darwin Correspondence Database.

Hooker, J. D. (1845a) Letter to Darwin. 5 July. Darwin Correspondence Database

Hooker, J. & Lyell, C. (1858) to Linnean Society of London. 30 June. Darwin Correspondence Database

Hooker, J. D. (1859) *Introductory essay to the Flora Tasmaniae: The botany of the Antarctic voyage of HM discovery ships Erebus and Terror in the years 1839-1843.* Vol. 3. London: Lovell Reeve.

Hooker, J. D. (1860) Review of Darwin's theory on the origin of Species by means of Natural Selection. *American Journal of Science and Arts.* Second series. pp.153-184.

Hooker, J. (1868) Dr Hooker's Inaugural Address Before the British Association for the Advancement of Science. In Little, E. (ed) *The Living Age*. Fourth Series. Volume 9. October-December. Boston: Little and Gay. pp.195-207.

Hooker.org: Hooker's Biography. 3. Imperial Botany

Hope, T. In three Volumes. (1831) An Essay on the Origin and Prospects of Man. Volume II. London. John Murray. Volume 8. Issue 1-4, 1866.

Howard, J. (1982) *Darwin*. Oxford. Oxford University Press.

Jackson, C. E. (1992) Prideaux John Selby: A Gentleman Naturalist. Christine E. Jackson. Northumberland. Spredden Press.

Jameson, W. (1853) Contributions to the history of the Relationship between Climate and Vegetation in the various parts of the Globe. On the Physical aspect of the Punjab its Agriculture and Botany. Journal of the Horticultural Society of London.Vol. 8. pp. 273-314.

Jeake, S. (1701) *A Compleat Body of Arithmetick*, in four books. London. Newborough.

Jeffrey, J. and Howie, C. (1879) *The Trees and Shrubs of Fife and Kinross*. Self-published. Private circulation.

Jenyns, L. (1885) *Reminiscences of Prideaux John Selby*. (Brief Notices of Some Other North Country Naturalists. Self-published. Private circulation book.

Johnson, C. W. (1841) On the Improvement of Peat Soils, Prize Essay. *Journal of the Royal Agricultural Society of England,* Volume 2. pp.390-399.

Johnson, C. W. (1842) Plantation. *The Farmer's Magazine* January to June. Vol. 5 pp. 364-368.

Johnson, D. R. (2010) *Nietzsche's Anti-Darwinism*. Cambridge. Cambridge University Press.

Johnson, M. (1993) *Williams Geology Newsletter* Vol. II. Bicentennial Edition. Summer.

Jones, E. J. M. (1992) Data Relative to the ManuKau-Waitemata Land Company. *Auckland Waikato Historical Journal*, September. No 61. pp.27-28.

Jones, E. J. M. (2000) An Historical Account of Matakana History by the Granddaughter of James Matthew, Mrs Errol Jones. Gourdiehill. Posted on the Denmylne blog. June 15th. 2012. http://denmylne.wordpress.com/2012/06/15/gourdiehill/. Also available at: Matakana Wharf reserve Management Plan: http://www.rodney.govt.nz/DistrictTownPlanning/plans/ReserveManage ment/Documents/Matakana_Wharf/MatakanaWharf_RMP_plan.pdf

Jones, E. J. M. (2010) *Shadows On My Wall, The Memoirs of Errol Jones.* Plymouth (New Zealand) PublishMe. ISBN 978-0-473-1644-9.

Jones, R. B. (2011) *20% Chance of Rain: Exploring the Concept of Risk.* John Wiley and Sons.

The Journal of Agriculture (1831) Vol. 3.

Journal of Horticulture, Cottage Gardener and Country Gentleman (1866) A Magazine of gardening, Rural and Domestic Economy, Botany and Natural History. Vol. 11 (new series). p.311.

Journal des Travaux (1839) de la Société Française de Statistique Universelle. Volume 10. Page 326

Judd, J. W. (1910) Darwin and Geology. In Seward, A. C. (ed.) *Darwin and Modern Science - Essays in Commemoration of the Centenary of the Birth of Charles Darwin and of the Fiftieth Anniversary of the Publication.* Cambridge. Cambridge University Press.

Judd, J. W. (1910) *The Coming of Evolution: The Story of the Great Revolution in Science.* Cambridge. Cambridge University Press.

Juengst, E. T. (2007) Population Genetic research and Screening: Conceptual and Ethical Issues. In Steinbock, B. (Ed.). *The Oxford Handbook of Bioethics.* Oxford. Oxford University Press.

Kelly, A. and Kelly, M. (2009) *Darwin for the love of science.* Bristol. Bristol Cultural Development Partnership.

Kempton, K. P. (1926) Harry in a Hurry. *Boys Life.* pp.70-71. (at p. 70).

Kew (2013b) Kew.org About Library and Art Archives:

Keynes, R. (2003) *Fossils, Finches, and Fuegians: Darwin's Adventures and Discoveries on the Beagle.* Oxford. Oxford University Press.

King, Fitzroy and Darwin, C. (1839) *Narrative of the Surveying Voyages of Her Majesty's Ships Adventure and Beagle.* Proceedings of the second expedition, 1831-1836, under the command of Captain Robert FitzRoy. Appendix to Volume II. London. Henry Colburn.

Kisia, S. M. (2011) *Vertebrates: Structures and Functions.* Abingdon. CRC Press. Taylor and Francis Group.

Knapp, S., Sanders, L. and Baker, W. (2002) *Alfred Russel Wallace and the Palms of Amazon.* Palms. Vol. 46. 3. pp.109-199.

Knapton, S. (2014) Did Charles Darwin 'borrow' the theory of natural selection? *Telegraph.* Online Wednesday May 28. www.telegraph.co.uk/news/science/science-news/10859281/Did-Charles-Darwin-borrow-the-theory-of-natural-selection.html

Knight, T. A. (1797) *A Treatise on the Culture of the Apple & Pear and on the Manufacture of Cider & Perry.* Ludlow H. Procter.

Knox, J. (1850) *Map of the Basin of the Tay, including the greater part of Perth Shire, Strathmore and the Braes of Angus or Forfar.* Edinburgh. W. & A. K. Johnson.

Koonin, E. V. (2011) *The Logic of Chance: The Nature and Origin of Biological Evolution.* FT Press.

Kourilsky, P. (2012) Selfish cellular networks and the evolution of complex organisms. *Comptes Rendus Biologies.*

Kroon, A. M. and Saccone, C. (1974) *The biogenesis of mitochondria: transcriptional, translational and genetic aspects.* New York: Academic Press, 1974.

Kottler, M. J. (1974) Alfred Russel Wallace, the Origin of Man, and Spiritualism. *Journal of the History of Science Society.* Vol. 65, No.2. June. pp.144-192.

Kuhn, T. S. (1962) *The Structure of Scientific Revolutions.* (second edition, enlarged). Chicago. University of Chicago Press.

Laboratory Practice (1951). London. Trade Press. pp.231-232.

Laland, K. N. and Brown, G. (2011) *Sense and Nonsense: Evolutionary Perspectives on Human Behaviour.* Oxford. Oxford University Press.

Langlois, R. N. and Everett, M. J. (1994) In Magnusson, L. (ed) Evolutionary and Neo-Schumpeterian Approaches to Economics. *What is Evolutionary Economics?* Boston. Kluwer Academic Publishers.

Langlotz, D. D., Warme, B. (1996) *Study Guide to Accompany Sociology: a Window on the World.* Nelson Thomson Learning.

Lau, G. F. (2012) *Ancient Alterity in the Andes: A Recognition of Others.* Abingdon. Routledge.

Lawrence, W. (1819) Lectures on Physiology, Zoology, and the Natural History of Man. Delivered at The Royal College of Surgeons. London. J. Callow.

Lawson, C. (1999) Commons Contribution to Political Economy. In O'hara. P. A. *Encyclopaedia of Political Economy.* Vol. 1. London. Routledge.

Laycock, T. (1855) Further Researches into the Functions of the Brain. *British and Foreign Medico-chirurgical Review,* Volume 16. July to October. pp.155-187.

Lehman, M. M., Lazlo, A. and Belady (1985) *Program evolution: processes of software change.* London. Academic Press.

Leidy, J. (1858) *The Journal of the Louisiana State Medical Society.* Volume 15, page 677.

Leff, A. (2003) Thomas Laycock and the romantic genesis of the cerebral reflex ACNR. *History of Neuroscience.* Volume 3 Number 1. March/April. pp.26-27.

Leighton, W. A. (1851) *Report of the Council of the Ray Society* In: The British species of angiocarpous lichens, elucidated by their sporidia. Page 179.

Leland, T. (1764) *A Dissertation on the Principles of Human Eloquence with Particular Regard to the Style and Composition of the New Testament*, in which the Observations on this Subject by the Lord Bishop of Gloucester in his Discourse on the Doctrine of Grace are distinctly considered being the substance of several lectures read in the Oratory School of Trinity College Dublin. Ludgate Street. W. Johnston.

Leonard, R. D. and Jones, G. T. (2002) Natural Selection. In Hart, J. P. and Terrell, J. (eds) (2002) *Darwin and Archaeology: A Handbook of Key Concepts.* Westport. Greenwood Publishing.

Léopold, G. et al. (1840) *Cuvier's Animal Kingdom*: Arranged According to Its Organisation; Forming the Basis for a Natural History of Animals, and an Introduction to Comparative Anatomy. Mammalia, Birds, and Reptiles, by Edward Blyth. The Fishes and Radiata, by Robert Mudie. The Molluscous Animals, by George Johnston, The Articulated Animals, by J.O. Westwood, Illustrated by Three Hundred Engravings on Wood. W.S. Orr. London.

Leslie, J. (1823) *Elements of Natural Philosophy.* Volume 1. Edinburgh. W and C Tait.

Levine, G. (2011) *Darwin the Writer.* Oxford. Oxford University Press.

Lewis, J. H. (2016) *Origins of Charles Darwin evolutionary theory challenged by Nottingham academic.* Nottingham Post. March 27th.

Lewis, R. et al. (2002) *Life.* Boston. McGraw-Hill.

Lhwyd, E. (1712) In Londres et al. (eds) *Philosophical Transactions, Giving Some Acconpt of the Present Undertakings, Studies and Labors of the Ingenious in Many Considerable Parts of the World* Volume 28. London. See page 506.

Library of Congress (1840) Catalogue of the Library of Congress, in the capitol of the United States of America. December 1839 p.127 Washington.

The London Medical Gazette (1837). (Anon) Vol. 20. Vol. II for the session 1836-37. Review of medical essays by B. J. Hungerford Sealy. pp.313-.314.

Lorimer, P. (1857) *Precursors of Knox, or Memoirs of Patrick Hamilton.* Edinburgh. Thomas Constable and Co.

Lossing, B. J. (1855) *Our Countrymen: or brief memoirs of eminent Americans.* New York. Ensign Bridgeman and Fanning.

Louçã, F. and Perlman, M. (2000) *Is economics an evolutionary science?: The legacy of Thorstein Veblen.* Cheltenham (UK). Northampton (Mass.) E. Elgar.

Loudon, J. C. (1822) *An Encyclopaedia of Gardening,* Comprising the Theory and Practice of Horticulture, Floriculture, Arboriculture and Landscape-gardening, Including. a General History of Gardening in All Countries. London. Longman, Hurst, Rees, Orme and Brown.

Loudon, J. C. (1831) *An Encyclopædia of Agriculture*: Comprising the Theory and Practice of the Valuation, Transfer, Laying Out, Improvement, and Management of Landed Property; and the Cultivation and Economy of the Animal and Vegetable Productions of Agriculture, Including All the Latest Improvements; a General History of Agriculture in All Countries; and a Statistical View of Its Present State, with Suggestions for Its Future Progress in the British Isles. London: Longman, Rees, Orme, Brown, and Green.

Loudon, J. C. (1832) Matthew Patrick On Naval Timber and Arboriculture with Critical Notes on Authors who have recently treated the Subject of Planting. *Gardener's Magazine.* Vol. VIII. p.703.

Loudon, J. C (1835) An Encyclopaedia of Gardening: Comprising the Theory and Practice of Horticulture, Floriculture, Arboriculture and Landscape- *Gardening.* London. Longman, Orme, Brown, Green, Longman. Loudon, J. C. (1838) *Arboretum et Fruticetum Britannicum* (1838). Or the Trees and Shrubs of Britain. Pictorially and Botanically Delineated. In Eight Volumes. Vol 1. London. Longman, Orme, Brown, Green and Longman.

Loudon, J. C. (1850) *Loudon's Hortus Britannicus A Catalogue of All the Plants, Indigenous, Cultivated in, or Introduced to Britain.* Part 1. London.

Lucretious (50 BCE) *On the Nature of Things.* Translated by William Ellery Leonard (1916). Dutton. Available free from American Libraries Internet Archive:

Low, D. (1834) *Elements of Practical Agriculture.* Edinburgh. Bell and Bradfute.

Low, D. (1842) *On landed property, and the economy of estates: comprehending the relation of landlord and tenant, and the principles and forms of leases--farm buildings, enclosures, drains, embankments, roads, and other rural works - minerals - and woods.* London. Longman, Brown, Green and Longmans.

Lurie, M. (1969) The Darwinian selection theory of antibody formation. *Journal of Theoretical Biology.* Volume 23, Issue 3, June 1969, pp.380–386.

Lyell, C. (1830) *Principles of Geology: being an attempt to explain the former changes of the Earth's surface.* By reference to causes now in operation. In two volumes. Vol. 1. London. John Murray.

Lyell, C. (1832) *Principles of Geology: being an attempt to explain the former changes of the Earth's surface.* By reference to causes now in operation. In two volumes. Vol 2. London. John Murray. Charles Lyell in New York State. Geological Society, London, Special Publications. Vol.143. pp.71-81

Lyell, C. (1868) *Principles of Geology: being an attempt to explain the former changes of the Earth's surface.* By reference to causes now in operation. In two volumes. Vol. 2. Tenth and Entirely Revised Edition. London. John Murray.

Lyell, C. (1881) *Life, letters and journals of Sir Charles Lyell* (Baronet). London. J. Murray.

Lyell, K. M. (ed) (2010) *Life, letters and journals of Sir Charles Lyell, Bart.* Volume 2. Cambridge. Cambridge University Press.

McCalman (2009) *A Lunatic Idea: British Science and Evolution on the Eve of Darwin's Origin of Species.* Paper presented Darwin symposium, National Museum of Australia, 26 February 2009.

Machinery and production engineering (1920) Volume 15. p.366.

Magazine of Natural History and Journal of Zoology, Botany, Mineralogy, Geology and Meteorology (1831). Vol. IV. p.571.

Malthus, T. (1798) *An Essay on the Principle of Population, as it Affects the Future Improvement of Society, with Remarks on the Speculations of Mr. Godwin, M. Condorcet, and Other Writers.* Dusseldorf; Darmstadt: Verl. Wirtschaft und Finanzen.

Malthus, T. (1803) *An essay on the principle of population: or, a view of its past and present effects on human happiness.* London. J. Johnson.

Manier, E. (1978). *The young Darwin and his cultural circle: a study of influences. which helped shape the language and logic of the first drafts of the theory of natural selection.* Dordrecht. Reidel.

Manuel, D. E. (1996) Marshall Hall (1790-1857) *Science and Medicine in Early Victorian Society.* Amsterdam. Editions Radopi.

Maranda, P. (1972) (ed) *Mythology.* Harmondsworth. Penguin.

Marchant, J. (ed.) (1916) *Alfred Russel Wallace Letters and Reminiscences.* Volume 2. London: Cassell.

Main, J. (1835) *Illustrations of Vegetable Physiology, Practically Applied to the Cultivation of the Garden, The Field, and the Forest.* London. Orr and Smith.

Malec, G. (2015) There Is No Darwin's Greatest Secret. *Filozoficzne Aspekty Genezy.* Vol 12. pp.325-331.

Markel, A. L. and Trut, L. N. (2011) Behaviour, Stress and Evolution in Light of the Novosibirsk Selection Experiments. In Gissis, S. B. and Jablonka, E. (eds) *Transformations of Lamarckism: From Subtle Fluids to Molecular Biology.* Cambridge. Mass. MIT Press.

Martin, M. (2001) *A Long Look at Nature: The North Carolina State Museum of Natural Sciences.* University of North Carolina Press. p.41.

Martinez, A. A. (2011) *Science Secrets: The Truth about Darwin's Finches, Einstein's Wife, and Other Myths.* Pittsburgh. University of Pittsburgh Press.

Marx, W. and Bornmann, L. (2013) Tracing the origin of a scientific legend by Reference Publication Year Spectroscopy (RPYS): the legend of the Darwin finches. In *Scientometrics.* October 6th.

Masur, L. P. (2001) 1831 *Year of the Eclipse.* New York. Hill and Wang.

Matthew, P. (1829) Some Account of the Fruits grown in Gourdie Hill Orchard Carse of Gowrie with Remarks. In a Letter from Patrick Matthew Esq. to the Secretary dated 3 December 1827. *Memoirs of the Caledonian Horticultural Society.* Fourth Volume. Edinburgh. Maclachlan and Stewart. London Simpkin and Marshall.

Matthew, P. (1831) *On Naval Timber and Arboriculture; With a critical note on authors who have recently treated the subject of planting.* Edinburgh. Adam Black.

Matthew. P. (1832) On Pruning. *The Quarterly Journal of Agriculture.* Volume 3. February to September. pp.300-308.

Matthew, P. (1839) *Emigration fields: North America, the Cape, Australia, and New Zealand;* describing these countries, and giving a comparative view of the advantages they present to British settlers. Edinburgh. Adam and Charles Black.

Matthew, P. (1839a) *Prospectus of the Scots New Zealand Land Company.* Edinburgh. Adam & Charles Black.

Matthew, P. (1839b) *Two addresses to the men of Perthshire and Fifeshire: containing propositions of a plan of national education, and other social improvements and reforms.* Edinburgh. A and C Black.

Matthew, P. (1860) Letter to the Gardeners Chronicle. Nature's law of selection. *Gardeners' Chronicle and Agricultural Gazette* (7 April): pp.312-13.

Matthew, P. (1860a) Origin of Species. *Farmers Magazine.* Vol. 18 (third edition) July-December. pp.30-32. (Letter dated May 19th).

Matthew, P. (1860b) Letter to the Gardeners Chronicle. Nature's law of selection. *Gardeners' Chronicle and Agricultural Gazette* (12 May) p.433.

Matthew, P. (1860c) How to Grow Red Clover on Clay Land. *The Farmers Magazine.* pp.406-407 Vol. 18 (third series) January to July (letter dated March 26th 1860).

Matthew, P. (1861) The Greatest National Evil. The System of Land Occupancy in Britain Incompatible With Improvement. *The Farmer's Journal.* March 12. pp.388–391.

Matthew, P. (1861a) The Potato Blight and Harvest Prospects in the North. *The British Farmer's Magazine.* Volume 41. New Series. p.90.

Matthew, P. (1862) Protection to Property. *The Farmer's Magazine.* Vol. 22 (third series) July-December, pp.412-413.

Matthew, P. (1862a) Letter: Matthew, Patrick to Darwin, C. R. December 3rd. Darwin Correspondence Database.

Matthew, P. (1864) *Schleswig-Holstein.* London. Spottiswoode.

Matthew, P. (1867) Letter in the Dundee Advertiser. In Dempster, W. J. (1996) *Evolutionary Concepts in the Nineteenth Century.* Edinburgh. The Pentland Press.

Matthew, P. (1870a) Letter to the Dundee Advertiser. 11th February. Available on pp.152-154 in Dempster (1983).

Matthew, P. (1870b) Letter to the Dundee Advertiser. 4th January. Available on pp.142-145 in Dempster (1983).

Matthew, P. (1870c) Letter to the Dundee Advertiser. 21st January. Available on pp.146-149 in Dempster (1983).

Matthew, P. (1871) Letter: Matthew, Patrick to Darwin, C. R. 12 Mar. Darwin Correspondence Database.

May, W. (1911) *Zoologische Annalen: Zeitschrift für Geschichte der Zoologie* Volumes 3-4. pp.280-295.

Mayr, E. (1982) *The growth of biological thought: diversity, evolution, and inheritance.* Cambridge, Mass. Harvard University Press.

McCalman (2009) *A Lunatic Idea: British Science and Evolution on the Eve of Darwin's Origin of Species.* Paper presented Darwin symposium, National Museum of Australia, 26 February 2009.

McClellan III, J. E. and Dorn, H. (2006) *Science and Technology in World History: An Introduction.* Baltimore. John Hopkins University Press.

McGrath (2006) *A Scientific Theology.* London. T & T Clark.

McShea, D. W. and Rosenberg, A. (2008) *Philosophy of Biology: A Contemporary Introduction.* Routledge.

Mechanics' Magazine and Register of Inventions and Improvements (1835) Volume 5. New York. Minor, D. K.

Merton, R. K. (1957) Priorities in Scientific Discovery: A Chapter in the Sociology of Science. *American Sociological Review.* Volume 22. No. 6. December. pp. 635-659.

Merton, R. K. (1968) The Matthew Effect in Science: The reward and communications systems of science are considered. *Science.* 159 (3810) pp.56-63. January 5.

Merton, R. K. (1948) The self-fulfilling prophecy. *The Antioch Review*, 8, 1948. pp.193-210.

Merton, R. K. (1973) *The Sociology of Science: Theoretical and Empirical Investigations*. Chicago. University of Chicago Press.

Meteyard (1871) *A Group of Englishmen: Being records of the younger Wedgwood's and their friends*. London. Longmans, Green and Co.

The Metropolitan (1831). Vols. 2-3 p.44.

Millhauser, M. (1959) *Just Before Darwin: Robert Chambers and the Vestiges*. Middletown, Conn. Wesleyan University Press.

Molyneux, T. (1695) Some Notes upon the foregoing Account of the Giants Causeway serving to further illustrate the same. In: *Philosophical Transactions, Giving Some Account of the Present Undertakings, Studies and Labors of the Ingenious in Many Considerable Parts of the World*. Vol.18 for the year 1694. London. The Royal Society.

Mondobbo, J. B. (Lord) (1774) Lord Monboddo, Orangutans and the Origins of Human Nature. Volume 6 of *Animal rights and souls in the eighteenth century*. Bristol. Thoemmes Press.

Monthly Review, or *Literary Journal* (1789) January to June.

Moore, J. M. (1981) *The post-Darwinian controversies: a study of the Protestant struggle to come to terms with Darwin in Great Britain and America: 1870-1900*. Cambridge. Cambridge University Press.

More, H. (1653) *Conjectura Cabbalistica*: or, a conjectural essay of interpreting the minde of Moses according to a three-fold Cabbala, viz.literal. philosophical. mystical, or, divinely moral. The defence of the threefold Cabbala. London. J. Flesher.

Morgan, S. (1835) *The Princess, or The Beguine*. Paris. Boudry.

Mure, W. (1854) *A critical history of the language and literature of ancient Greece*, Volume 3. London. Longman and Co.

Murphy, E. (1834) *Irish Farmer's and Gardener's Magazine and Register of Rural Affairs*. Volume 1.

Nail, S. (2008) *Forest Policies and Social Change in England*. London. Springer.

Napier, J. (2011) *Darwin and Evolution: Bullet Guide*. Hachette.

Neate, G. (1985) *MEMEX, Evaluation of a Search Engine*: Interim Report to British Library Research and Development on Project SI/G/627.

Nelson, R. W. (2009) *Darwin Then and Now: The Most Amazing Story in the History of the Universe*. New York. Bloomington. iUniverse.

Netter, F. H. (1948) *The Ciba collection of medical illustrations; a compilation of pathological and anatomical paintings prepared*. Volume 4. Summit, N.J., Ciba Pharmaceutical Products. p. 111.

The New England Farmer and Horticultural Journal (1833) Vol. 2. p.391.

New Monthly Literary Journal (1833) Vol. 2. July to December. p.139.

New Zealand Journal (1843) Captain Fitzroy, The New Governor. Vol. IV. pp. 98 and 221.

New Zealand Journal (1846) The New Zealand Journal. List of Publications Relevant to New Zealand. Vol. VI p.93.

Newman, E. (1860) *The Zoologist*. Vol. 19. page 7600.

Nicholson, W. (1790) *An introduction to Natural Philosophy: Illustrated with copper plates*. Volume 2. London. J. Johnson.

Nickles, T. (2009) The Strange Story of Scientific method. In Meheus, J. and Nickles, T. (eds.) *Models of Discovery and Creativity*. London. Springer.

Nicol, W. (1802) *The Forcing Fruit and Kitchen Gardener*. Edinburgh. William Creech.

Nishiura, H. (2007) Discussion: Emergence of the concept of the basic reproduction number from mathematical demography. *Journal of Theoretical Biology* 244: pp.357-364.

Nixon, R. (2012) *Neoliberalism, Genre, and The Tragedy of the Commons*. PMLA Volume 127, Number 3, May 2012, pp.593–599 (7).

Nolan, S. (2013) He invented 'origin of species' and warned that species are dying out: Now 'forgotten father' of evolution Alfred Russel Wallace is celebrated at last. *Mail Online* – Science and Tech. 23rd January:

Norman, R. (2004) *On Humanism*. London. Routledge.

Nunberg, G. (2008) *All Thumbs*. Berkeley:

Okasha, S. (2002) *Philosophy of Science: A Very Short Introduction*. Oxford. Oxford University Press.

Oliver, H. (2008) *March Hares and Monkeys' Uncles: Origins of the Words and Phrases we use in Everyday Life*. London. Metro Publishing.

Orlove, M. J. (1975). A model of kin selection not invoking coefficients of relationship. *Journal of Theoretical Biology*, Volume 49, Issue 2, February 1975, pp.289–310.

Otto, S. L. (2011) *Fool Me Twice: Fighting the Assault on Science in America*. New York, Rodale: Distributed to the trade by Macmillan.

Oughtred, W. (1694) *Key of the Mathematics*. London. Saulsbury.

Owen, R. (1841) On British Fossil Reptiles. Part II. From the *Report of the British Association for the Advancement of Science for 1841*. London.

The Oxford Library of Words and Phrases (1990) Vol. 1. *The Oxford Dictionary of Quotations* (second edition) London. Guild Publishing.

Palmer, T. (2003) *Perilous Planet Earth: Catastrophes and Catastrophism Through the Ages*. Cambridge. Cambridge University Press.

Papworth, W. (1858) Notes on the Assumed use of Chestnut Timber in the Carpentry of Old Buildings. *The Civil Engineer and Architects Journal*. Vol. 21. p.295.

Papworth, W. (1858) Notes on the Assumed Use of Chestnut Timber in the Carpentry of Old Buildings, By Wyatt Papworth Architect Read at the Ordinary General Meeting of the Royal Institute of British Architects June 14 1858. In *Papers Read at the Royal Institute of British Architects*.

Paris, J. A. (1825) *Pharmacologia*. 6th edition. Vol. 1. London. W Phillips.

Pardy, D. (2012) *Introducing Leadership*. Oxford. Elsevier.

Partington, C. F. (ed.) (1835) *The British Cyclopædia of Natural History: combining a scientific classification of animals, plants, and minerals.* (Vol.1). London: W. Orr & Smith.

Partington, C. F. (ed.) (1837) *The British Cyclopædia of natural history: combining a scientific classification of animals, plants, and minerals.* (Vol.3) London: W. Orr and Co.

Partington, C. F. (ed.) (1838) *The British Cyclopædia of Natural History: combining a scientific classification of animals, plants, and minerals.* (Vol.6). London: W. Orr & Smith.

Pasternak, C. (2007) *What makes us human?* Oxford. One World.

Pearl, R. (1941) Some Biological Considerations About War. *The American Journal of Sociology.* Vol. 46, No. 4. January. pp.487-503.

Pearn, A. (2009) (ed) *A Voyage Round the World: Charles Darwin and the Beagle Collections in the University of Cambridge.* Cambridge. Cambridge University Press.

Pearrson, K. A. (2012) *Viroid Life: Perspectives on Nietzsche and the Transhuman Condition.* Routledge.

Pepys, S. (1665) *The Diary of Samuel Pepys*, Volume 2. Random House. New York.

Petroski, H. (2010) *Images of Progress: Conferences of Engineers.* In Bryson, B. (editor) Seeing Further: The Story of Science and the Royal Society. London Harper Collins.

Phillips, M. L. (1841) *Introductory Lecture to the Courses on Physical Science. Manchester New College.* London. Simpkin, Marshall and Co.

Pinsdorf, M. K. (1997) Engineering Dreams Into Disaster: History of the Tay Bridge. *Business and Economic History.* Volume 26. Number 3. Winter. pp. 491-504.

Playboy Magazine, (1969) "The Playboy Interview: Marshall McLuhan," March. pp.26-27, 45, 55-56, 61, 63.

Podolefsky, A. and Brown, P. J. (2007) *Applying anthropology: an introductory reader.* Boston : McGraw-Hill.

Pólya, G. (1954) *Mathematics and Plausible Reasoning: Induction and analogy in mathematics.* Princeton. Princeton University Press.

Popper, K. R. (1959) *The Logic of Scientific Discovery.* London. Routledge.

Porter, D. M. (2012) Why Did Wallace Write To Darwin? *The Linnean.* Volume 28 (1) pp.17-25.

The Postal Record (1908). National Association of Letter Carriers. Volumes 21-22. Page 27.

Potter, J. and Wetherell, M. (1987) *Discourse and Social Psychology.* London. Sage.

Powell, B. (1855) *Essays on the spirit of the inductive philosophy: the unity of worlds and the philosophy of creation.* London. Longman, Brown, Green, and Longmans.

Powell, B. (1858) Christianity without Judaism. *British and Foreign Evangelical Review.* Vol. VII. July. Essay written 1857. pp.485-522.

Preston, W. (1803) *THE ARGONAUTICS OF APOLLONIUS RHODIUS, TRANSLATED INTO ENGLISH VERSE. WITH NOTES CRITICAL, HISTORICAL, AND EXPLANATORY, AND DISSERTATIONS.* Vol. III. Dublin Graisberry and Campbell.

Priestley, Lady (1908) *The Story of a Lifetime.* London. Kegan Paul, Trench, Trubner and Co.

Pritzker, S. R. and Runco, M. A. (1999) *Encyclopedia of creativity* Vol. 1. A H. San Diego, Calif. Academic Press.

Provine, W. B. (1989) *Sewall Wright and Evolutionary Biology.* Chicago. University of Chicago Press.

Pullman, J. M. (1805) *Short Stories in the Life and Teachings of Jesus. From B.C. 5 to A.D. 30. For Sunday Schools. XLI. Our Duties to God and Man.* Lynn. Mass. The Nichols Press.

Purves, G. (1818) Gray versus Malthus. *The Principles of Population and Production.* London. Longman, Hurst, Rees, Orme and Brown.

Quarterly Educational Magazine and Record of the Home and Colonial School Society. (1848) Vol. 1. page 363.

The Quarterly Literary Advertiser (1831). Advertisement for On Naval Timber and Arboriculture. January to October. p.393.

The Quarterly Review (1833) April to July. Vol. 49. (pp.126-127). London. John Murray.

Radwan☐ ski, P. A. (1966) *Man, the known*. New York, Universum Press Co.

Rafinesque, C. S. (1836) *New Flora and Botany of North America*. Philadelphia. Self-published.

Rafinesque, C. S. (1836a) *Flora Telluriana*. First Part. Philadalphia.

Rampino, M. R. (2011) Darwin's error? Patrick Matthew and the catastrophic nature of the geologic record. *Historical Biology: An International Journal of Paleobiology*. Volume 23, Issue 2-3.

Rana, L. (2008) *Geographical Thought: A Systematic Record of Evolution*. New Delhi. Ashok Kumar Mittal.

Reich, C. G. (1800) Professor Reich's Theory and Practice of Fever. In Bradley, T., Baty, R., and Noeden, A. A. (Eds) *The Medical and Physical Journal*. Vol. IV. June to December. pp.556-563.

Rennison. N. (2007) *Peter Mark Roget: The Man Who Became a Book*. Harpendon. Pocket Essentials.

Report of the Committee of the Berwick and Kelso Railway Committee (1837). Edinburgh. See page 5.

Richards, A. (2012) *Slow Dorset*. Guilford, Connecticut. The Gobe Pequot Press.

Roberts, M. B. V. (1986) *Biology: a functional approach*. (4th Edition) Cheltenham. Thomas Nelson and Sons.

Roger, C. (1849) *History of St Andrews*. Edinburgh. Adam and Charles Black.

Roget, P. M. (1834) *The Bridgewater Treatise on the Power, Wisdom and Goodness of God As Manifested in the Creation. Treatise V. Animal and Vegetable Physiology*.

Rose, M. R. (2000) *Darwin's Spectre: Evolutionary Biology in the Modern World*. Princeton University Press.

Romanes, G. J. (1886) Physiological selection. An additional suggestion on the origin of species. *Nature* 34. pp.314-316, 336-340 and 362-365.

Rose, R. (2009) *For all the Tea in China: Espionage, Empire and the secret formula of the world's favourite drink*. London. Huchinson.

Rothery, H. C. (1880) *Tay Bridge Disaster, Report of the Court of Inquiry, and Report of Mr. Rothery, Upon the Circumstances attending the Fall of a Portion of the Tay Bridge on the 28th December 1879*. London. Board of Trade.

Rowe, E. R. (1855) *My Life: Or, The autobiography of a village curate*. London. Henry Vizetrelly.

Royal Botanic Gardens, Kew (1899) *Bulletin of miscellaneous information*. Royal Gardens, Kew. (1899). Kew. Surrey. Royal Botanic Gardens.

Royal Society (1840) *Philosophical Transactions of the Royal Society of London*, Volumes 111-120. 1833-1840. London. Richard Taylor.

Rudwick, M. J. (1988) *The Great Devonian Controversy: The Shaping of Scientific Knowledge Among Gentlemanly Specialists*. Chicago. University of Chicago Press.

Ruiz-Mirazo, K., Umerez, J. and Moreno, A. (2008) Enabling conditions for open-ended revolution. *Biology & Philosophy*. January. Volume 23, Issue 1, pp.67-85.

Ruse, M. (1993, p. 19) *The Darwinian Paradigm: Essays on Its History, Philosophy, and Religious Implications*. New York. Routledge.

Ruse, M. (1995, pp.20-21) *Evolutionary Naturalism: Selected Essays*. Routledge.

Ruse, M. (1999) *The Darwinian Revolution: Science red in tooth and claw*. Chicago. University of Chicago Press.

Rutherford, M. and Samuels, W. J. (1996) *John R. Commons. Selected Essays*. Volume One and Two. London. Routledge.

Ryan, M. D. (ed.) (1833) *The London Medical and Surgical Journal*. London. Renshaw and Rush.

Salesa, D. L. (2011) *Racial Crossings: Race, Intermarriage, and the Victorian British Empire*. Oxford. Oxford University Press.

Secord, J. A. (1991) Edinburgh Lamarckians: Robert Jameson and Robert E. Grant. *Journal of the History of Biology*. Vol. 24. No. 1. pp.1-18.

Secord. J. A. (2000) *Victorian Sensation: The Extraordinary Reception, and Secret Authorship of Vestiges of the Natural History of Creation*. Chicago and London. The University of Chicago Press.

Selby, P. J. (1842) *A history of British forest-trees: indigenous and introduced*. London. van Voorst.

Selsam, H., Martel, H. (1973) *Reader in Marxist Philosophy*. New York: International Publishers.

Seward, A. C. (1909) *Darwin and Modern Science: Essays in Commemoration of the Centenary of the birth of Charles Darwin and of the 50th Anniversary of the publication of the Origin of Species*. Cambridge. Cambridge University Press.

Seward, A. C. (1912) Sir Joseph Hooker and Charles Darwin: The History of a Forty Year's Friendship. *New Phytologist*. Volume 11. Issue 5-6.

Shackle, G. L. S. (1961) *Decision, Order and Time in Human Affairs*. Cambridge: Cambridge University Press.

Shackle, G. L. S. (1983) The Bounds of Unknowledge. In J. Wiseman (ed), *Beyond positive economics*. New York: St. Martin's Press.

Shackle, G. L. S (1973) *Epistemics and Economics: A Critique of Economic Doctrines*. Cambridge. Cambridge University Press.

Schaeffer, F. (2010) *Patience with God: Faith for People Who Don't Like Religion (or Atheism)* Da Capo Press.

Shapiro, J. A. (1997) A Third Way. *Boston Review*. February/ March. p.6.

Schoeps, H. J., Winston, R. and Winston, C. (1968) *The religions of mankind*. Garden City, NY: Anchor Books. Doubleday.

Shermer, M. (2002) *In Darwin's Shadow: The Life and Science of Alfred Russel Wallace: A Biographical Study on the Psychology of History*. Oxford. Oxford University Press.

Sherman, M. and Sherman, B. (1963) It's a Small World. In: *Walt Disney's It's A Small World.* Wonderland Music Co. Walt Disney Productions. New York. Golden Books.

Simmonds, P. L. (ed) (1844) In New Zealand. *Simmonds Colonial Magazine and Foreign Miscellany.* January-April. Vol. 1. London. Foreign and Colonial Office.

Sinclair, J. (1793) *The Statistical Account of Scotland.* Vol. 9. Edinburgh. William Creech.

Slater, P. J. B. (2009) Darwin in Scotland: Professor P J B Slater on the Scottish connections of the birthday boy. *Scottish Review.* The Travel Review IV. 21st July Issue no 118.

Smith, S. (1827) Dr Southwood Smith's Lectures on Comparative and Human Physiology. *The London Magazine.* January to April. Vol. VII (new series).

Smith, A. (1860) Address of the President of the Royal Geological Society of Cornwall. Forty Seventh Annual Report of the Council. With the presidents Address, and Papers and Notices Read to the Society. Penzance Vibert. In Royal Geological Society of Cornwall. *Annual Report of the Council, with the President's Address.* Volumes 28-50.

Stott, R. (2012) *Darwin's Ghosts: In Search of the First Evolutionists.* London. Bloomsbury.

Strickland, H. E. (1849), Letter to Darwin. 31st January. Darwin Correspondence Project.

Strivens M (2003) The Role of the Priority Rule in Science. *Journal of Philosophy.* 100(55): pp.1- 33.

Sulloway, F. (1982) Darwin's Conversion: The Beagle Voyage and its Aftermath. *Journal of the History of Biology.* vol.15. pp.325-397.

Sulloway, F. (1984) Darwin and the Galapagos. *Biological Journal of the Linnean Society.* Vol. 2. pp.29-59.

Sutton, M. (2012) Expert Skeptics Suckered Again: Incredibly, the Famous Semmelweis Story is Another Supermyth. *BestThinking.com*. http://www.bestthinking.com/articles/science/biology_and_nature/bacteriology/expert-skeptics-suckered-again-incredibly-the-famous-semmelweis-story-is-another-supermyth

Sutton, M. (2012a) The Spinach, Popeye, Iron, Decimal Error Myth is Finally Busted. *BestThinking.com*. http://www.bestthinking.com/articles/science/chemistry/biochemistry/the-spinach-popeye-iron-decimal-error-myth-is-finally-busted

Sutton, M. and Hodgson, P. (2013) The Problem of Zombie Cops in Voodoo Criminology: Arresting the Police Patrol 100 Yard Myth. *Internet Journal of Criminology*.

Sutton, M. (2013) Fame at Last: Google Finds the Original Google after 110 Years in a Library 'Wilderness'. Criminology: the blog of Mike Sutton. *BestThinking.com:* http://www.bestthinking.com/thinkers/science/social_sciences/sociology/mike-sutton?tab=blog&blogpostid=20445%2c20445

Sutton, M. (2013a) 'Internet Dating for Nerds: Discovering Your Roots with Google'. Criminology: The Blog of Mike Sutton. *BestThinking.com*. http://www.bestthinking.com/thinkers/science/social_sciences/sociology/mike-sutton?tab=blog&blogpostid=20564

Sutton, M. (2013b) 'The Selfish Gene Myth is Bust: Richard Dawkins is an Invented Originator'. Criminology: The Blog of Mike Sutton. March 5th. *Best Thinking.com*.

Sutton, M. (2013c) Sutton's Mythbusting Protest. Wikipedia Myth Number 28. The Humpty Dumpty Myth. *BestThinking.com*. http://www.bestthinking.com/thinkers/science/social_sciences/sociology/mike-sutton?tab=blog&blogpostid=21530,21530

Sutton, M. (2014a) *Nullius in Verba: Darwin's greatest secret* (1st edition). Extended and detailed 601 page e-book. USA. Thinker Books, Thinker Media.

Sutton, M. (2014b) *The hi-tech detection of Darwin's and Wallace's possible science fraud: Big data criminology re-writes the history of contested discovery.* Papers from the British Criminology Conference (peer reviewed). British Society of Criminology.

Sutton, M. (2015) On Knowledge Contamination: New Data Challenges Claims of Darwin's and Wallace's Independent Conceptions of Matthew's Prior-Published Hypothesis. *Filozoficzne Aspekty Genezy*. Vol. 12. pp.167-205.

Sutton, M. (2016) Darwin's Greatest Secret Exposed: Response to Grzegorz Malec's De Facto Fact Denying Review of My Book. *Filozoficzne Aspekty Genezy*. Vol. 13. pp.1-10.

Swackhamer, C. (1840) *The United States Magazine and Democratic Review*, Volume 7.

Sykes, G. and Matza, D. (1957) Techniques of neutralization: A theory of delinquency. *American Sociological Review* 22: pp. 664–670.

Taleb, N. N. (2007) *The Black Swan: The Impact of the Highly Improbable*. Allen Lane/Penguin Books.

Taylor, B. B. (1996) *Education and the law: a dictionary*. Santa Barbara, Calif: ABC-CLIO.

Taylor, J. (1655) *Eniautos: a course of sermons for all the Sundays of the year*. London. Richard Royston.

Tecumseh Fitch, W. (2010) *The Evolution of Language*. Cambridge. Cambridge University Press.

Tee, G. J. (1984) Review of Patrick Matthew and Natural Selection. By W. J. Dempster. University of Auckland. *New Zealand Journal of History*: http://www.nzjh.auckland.ac.nz/docs/1984/NZJH_18_1_06.pdf

Tefft, J.C. *The Christ Is Not a Person: The Evolution of Consciousness and the Destiny of Man*. Bloomington In. iUniverse.

Telegraph (2014). Darwin 'stole' theory of natural selection. *The Daily Telegraph*. 28 May, p.12.

Terrie, P. G. (1994) *Forever Wild: A Cultural History of Wilderness in the Adirondacks*. Syracuse, New York. Syracuse University Press. See page 173.

Thagard, P. (1992) *Conceptual Revolutions*. New Jersey. Princeton University Press.

Thomas, R. (1797) *The cause of truth, containing, besides a great variety of other matter, a refutation of errors in the political works of T. Paine, and other publications of a similar kind, in a series of letters.* Dundee. Printed by Colvill for the author.

Thompson, J. V. (1827) *Memoir on the Pentacrinus Europaeus.* Cork. King and Ridings.

Thompson, D. (2008) *Counterknowledge: How We Surrendered to Conspiracy Theories, Quack Medicine, Bogus Science and Fake History.* London. Atlantic.

Tiezzi, E. (2005) Ecosystems and Sustainable Development. V. Southampton, *UK Computational Mechanics Publications.* p.16.
Tokumei, R. (2011) *Monkeys on Our Backs: Why Conservatives and Liberals Are Both Wrong About Evolution.* John Hunt Publishing, 2011.

Tomar and Singh (2003) *Evolutionary Biology.* Meerut. Rastogi Publications.

Vanberg, V. J. (2002) *The Constitution of Markets: Essays in Political Economy.* London. Routledge.

Tool, M. R. (1988) *Evolutionary Economics: Foundations of Institutional Thought.* London. M. E. Sharpe.

Townsend, J. (1786) *A Dissertation on the Poor Laws, By a Well-Wisher to Mankind.* London.

Tyson, E. (1706) The Anatomy of a Male Opossum. In *The Philosophical Transactions.* London. The Royal Society.

Tyson, E. (1721) Observations on the Opossum. In Jones, H. (ed.) *Philosophical Transactions from 1700 to 1720.* London, G. Strahan. The Royal Society.

The United Service Journal and Naval and Military Magazine (1831a) Part II. On Naval Timber, pp.457-466.

Uglow, J. (2002) *The Lunar Men: The Friends who made the future.* London. Faber and Faber.

van Loon, H. W. (1942). *Van Loon's Lives.* New York, Simon and Schuster.

van Wyhe, J. (2002) (ed) The Complete Work of Charles Darwin Online.

van Wyhe, J. (2009) *Charles Darwin's Shorter Publications*. Cambridge. Cambridge University Press.

van Wyhe, J. (2011). Was Charles Darwin an Atheist? *The Public Domain Review*. 28th June.

van Wyhe, J. (2013) *Dispelling the Darkness: Voyage of the Malay Archipelago and the Discovery of Evolution by Wallace and Darwin.* (second edition). World Scientific.

von Humboldt, A (1822) political Essay on the Kingdom of New Spain, Longman and co.

von Sydow, M. (2012) *From Darwinian Metaphysics towards Understanding the Evolution of Evolutionary Mechanisms A Historical and Philosophical Analysis of Gene-Darwinism and Universal Darwinism.* Gottingen Universitatsverlag Gottingen c/o SUB Gottingen.

Wainwright, M. (2008) It's Not Darwin's or Wallace's Theory.
The Origin of Species Without Darwin and Wallace. Wainwrightscience Blog. Thursday, 24 July 2008. http://wainwrightscience.blogspot.co.uk/

Wainwright, M. (2008) Natural Selection: It's Not Darwin's (Or Wallace's) Theory. *Saudi Journal of Biological Sciences* 15 (1) pp.1-8.

Wainwright, M. (2011). Charles Darwin: Mycologist and Refuter of His Own Myth. *FUNGI* Volume 4:1 Winter. pp. 3-20.

Waite, A. E. (1893) *A New Light of Mysticism: Azoth*; Or, The Star in the East London. Theosophical Publishing Society.

Waldman, H. (1975) *World Encyclopaedia of Black Peoples Conspectus*. St. Clair Shores, Mich.: Scholarly Press.

Wallace, A. R. (1845) Letter to Bates. December 28th. Wallace Letters Online. Natural History Museum. Unique WCP identifier. pp.346.346

Wallace, A. R. (1855) On the law which has regulated the introduction of new species. *The Annals and Magazine of Natural History*. Series 2. 16. pp.184-196.

Wallace, A. R. (1858) Letter to Darwin and Hooker. 6th October. Darwin Correspondence Database.

Wallace, A. R. (1858a) 6 October. Letter to his mother Mary Ann Wallace (nee Greenell): Unique WCP identifier: WCP369.369. Wallace Letters Online. Natural History Museum.

Wallace, A. R. (1867) 1 October. Letter to Charles Darwin, Wallace Correspondent Project.

Wallace, A. R. (1869) *The Malay Archipelago*. New York: Harper & Brothers.
Wallace, A. R. (1869a) Letter to Darwin, C. R. 18th April. Darwin Correspondence Project.

Wallace, A. R. (1871) *Contributions to the Theory of Natural Selection. A Series of Essays*. New York. Macmillan and Co.

Wallace, A. R. (1879) Butler's "Evolution, Old and New." *Nature*. Volume 20 June 12th.

Wallace, A. R. (1879a) 9 May. Letter to Samuel Butler. Unique WCP identifier: WCP1586. Wallace Letters Online. Natural History Museum.

Wallace, A. R. (1905) *My Life: A Record of Events and Opinions*, Volume 1. Cambridge. Cambridge University Press. Note: Taken here from digitally printed version (2011), Cambridge University Press.

Walsh, D. M. (2012) The Struggle for Life and the Conditions of Existence: Two Interpretations of Darwinian Evolution. In Brinkworth, M. and Weinert, F. (eds) *Frontiers Collection: Evolution 2.0: Implications of Darwinism in Philosophy and the Social and Natural Sciences*. New York. Springer.

Warren, L. (1998) *Joseph Leidy: The Last Man Who Knew Everything*. Yale University Press.

Weale, M. E. (2015), 'Patrick Matthew's law of natural selection', <u>*Biological Journal of the Linnean Society*</u>, 115/4. pp.785–791.

Weibull, J. W. (1997) *Evolutionary game theory*. Cambridge Mass. MIT Press.

Wei, G. (2002) *An introduction to genetic engineering, life sciences and the law*. Singapore: Singapore University Press, National University of Singapore.

Weisberg, R. W. (2006) *Understanding Innovation in Problem Solving, Science, Invention, and the Arts.* New Jersey. John Wiley & Sons.

Wells, W. C. (1818) *Two Essays: One Upon Single Vision with two eyes; The other On Dew. A Letter To The Right Hon. Lloyd, Lord Kenyon. And An Account of A Female of the White Race of Mankind, Part of Whose Skin Resembles that of a Negro; With Some Observations on the Causes of the Differences in Colour and Form Between the White and Negro Races of Man.* By the Late William Charles Wells. With a Memoir of his Life Written by Himself. London. Archibald Constable and Co. Edinburgh.

Wells, K. D. (1973) The Historical Context of Natural selection: The Case of Patrick Matthew. *Journal of the History of Biology.* Vol. 6. N0. 2. pp.225-258.

Westfahl, G. (1998) *The Mechanics of Wonder: The Creation of the Idea of Science Fiction.* Liverpool. Liverpool University Press.

The Westminster Review (1846) Affairs of New Zealand. Vol 46. American Edition. pp.70-117.

Whaley, P. (2002) *West Country History: Dorset.* Bristol. Venton.

Whewell, W. (1833) Astronomy and General Physics: Considered with reference to natural theology. *The Bridgewater Treatise III.* Second edition. London. William Pickering.

White, G., Blyth, E. and Mudie, R. (1850) *The natural history of Selborne, with its antiquities; naturalist's calendar, etc.* W. S. Orr. London.

Whitehurst, J. (1778) *An Inquiry into the Original State and Formation of the Earth.* London. Bent.

Wilkin, S. (Ed) (1835) *Sir Thomas Browne's Works.* Volume Three. London. William Pickering.

Wilson, J. Q. and Kelling, G. L. (1982) Broken Windows: The Police and Neighbourhood Safety. *Atlantic Monthly.* March.

The Windsor Magazine (1906), Vol. 24. p.547.

Williams, L. P. (1997) *Buffon: A Life in Natural History.* New York. Cornell University Press.

Wilner, E. (2006) Darwin's artificial selection as an experiment. Studies in History and Philosophy of Science Part C: *Studies in History and Philosophy of Biological and Biomedical Sciences*, Volume 37, Issue 1, March. pp.26–40

Wilson, D. S. (2007) *Evolution for Everyone: How Darwin's Theory Can Change the Way We Think About Our Lives*. New York, NY Delacorte Press.

Wilson, C. (1987) *Afterlife: an investigation of the evidence for life after death*. Garden City. N.Y. Doubleday.

Wilson, J. (1831) Essays on the Origin and Natural History of Domestic Animals. Essay IV On the Origin and Natural History of the Sheep and Goat. p. 355 *The Quarterly Journal of Agriculture*, Volume 2.

Wilson. J. (1837) The Elections, *Blackwood's Edinburgh Magazine*, Volume 42. pp.238-247.

Wilson, J. and Chambers, R. (1840) *The land of Burns: a series of landscapes and portraits*. Volume 1. Glasgow. Blackie and Son.

Wilson, J. M. (1848) *The Rural Cyclopedia*, or A general dictionary of agriculture: and of the arts, sciences, instruments, and practice, necessary to the farmer, stock farmer, gardener, forester, land steward, farrier. Edinburgh. A. Fullarton and Co.

Wilson, L. G. (editor), 1970. *Sir Charles Lyell's Scientific Journals on the Species Question*. New Haven and London: Yale University Press.

Wood, T. C. (2009) There is no Darwin Conspiracy. *Answers Research Journal* Vol. 2. pp.11-20.

Woodbury, L. (1832) Live Oak. *Report of the Secretary of the Navy*. December 15th. House of Representatives. Executive Documents. Duff Green. Washington.

Woodbury, L. (1833) Live Oak Timber for the Navy. *Military and Naval Magazine*. Vol. 1 Number 3.

Woodbury, L. (1838) Live Oak. House of Representatives. December 15, 1832. Report of the Secretary of the Navy on Live Oak. Navy Department. December 14th. In: *Register of Debates in Congress: Comprising the Leading Debates and Incidents of the Second Session of the Eighteenth Congress:* Dec. 6, 1824, to the First Session of the Twenty-fifth Congress, Oct. 16, 1837. Together with an Appendix, Containing the Most Important State Papers and Public Documents to which the Session Has Given Birth: to which are Added, the Laws Enacted During the Session, with a Copious Index to the Whole. Volume IX. Washington. (see p.128).

Woodbury, L. (1852) *Writings of Levi Woodbury, LL.D. Political, Judicial and Literary.* Volume 3 - Literary. Boston. Little, Brown and Company. p.361.

Wortley, R. (2011) *Psychological Criminology: An Integrative Approach.* Abingdon. Routledge.
Yankee Notions (1855) Volume 4, Issues 1-12. Page 167.

Yates, P. (1991) *The Cree Nurserymen of Lanarkshire and Surrey. Leicester.* The Cree Family History Society. http://www.cree.name/archives/nurserymen.pdf

Yeo, R. (1984) Science and Intellectual Authority in Mid-Nineteenth-Century Britain: Robert Chambers and Vestiges of the Natural History of Creation. *Victorian Studies*, Vol. 28. No.1. Autumn. pp.5-31. Indiana University Press.

Young (1785) National Improvements Upon Agriculture. In *Twenty-Seven Essays.* Edinburgh. John Bell.

Zimmer, C. (2003) *Evolution: the triumph of an idea.* London. Arrow Books, Random House.

Zirkle, C. (1941) Natural Selection before the 'Origin of Species'. *Proceedings of the American Philosophical Society.* Vol. 84. No. 1. (April 25) pp.71-123.

8 APPENDIX

On Naval Timber and Arboriculture: Additional Commentary by Mike Sutton

As disconfirming evidence for the Matthew Appendix Myth, that Matthew buried his entire hypotheses of the natural process of selection within a few scattered paragraphs/sentences/pages of an appendix to a book on an inappropriately unrelated obscure subject, this appendix exclusively presents text from Matthew's (1831) *NTA* in two parts.

Matthew draws very direct attention to his original ideas on natural selection. On page 2 of *NTA* he refers us to his Appendix A. On page 3, he refers us to his Appendix B. And on page 4, he refers us to Appendix C. So doing, at this initial point in his Introduction Matthew refers us very directly in this way so that we might here consider natural selection in the case of human "adaptation to condition" in the competitive struggle for existence between different peoples and of all humans over all other species. Contrary to the mere beliefs of biased *Darwin Cult* rhetoric, verifiable facts such as these reveal there is nothing limited, or at all obscure, or inappropriate, about natural selection in *NTA*.

Part One of this Appendix contains the text, directly relevant to Matthew's theory, which was included in the main body of his book. Part Two presents the relevant natural selection text from his appendix, which begins on page 363 of *NTA*.

In Part One, we can see how Matthew uses his concepts of struggle, competition, overtopping, artificial and natural selection, power of occupancy, circumstance, adaptability to condition and suitability to circumstance, all in the main body of his book along with his brief observations on climate and diversity and diversification of species. In Part Two, we can see why he concentrated those ideas in an appendix, because it is there that Matthew serves the Abrahamic "God" his redundancy notice.

Were his book to be deemed seditious and heretical, its appendix could be surgically removed.

The surgery could be performed by the publisher, bookseller or any owner.

In an age when both the church and scientific community ruled that natural theological explanations were not to be discussed by natural scientists, the structure of Matthew's book was one of rational and essential, contemporary, compromise. Contrary to modern Darwinist dysology that Matthew strangely hid his ideas in an appendix – an appendix was, in 1831, the very first place to look for radical ideas. Matthew's big idea was not buried anywhere, it was not placed in an appendix because he failed to understand its significance, but because he did. And so would those who read the heretical and seditious book, as pre-1858 reviews of it prove (e.g. Loudon 1832, United Services Journal 1831).

Most importantly then, it must be stressed that Matthew's discovery of the natural process of selection was not buried anywhere. There is no subtlety in *NTA*. The discovery of natural selection was purposefully concentrated and boldly placed in plain sight, where it could not possibly be overlooked. Matthew's hypothesis was deliberately concentrated in a highly visible, yet easily removed appendix. Because of its radical importance, Matthew's great discovery was read and its heresy was understood.

== Part One ==

From the main body of *NTA*

'NAVIGATION is of the first importance to the improvement and perfecting of the species in spreading, by emigration the superior varieties of man…'

Page 1

…an overflowing population, chained, from the state of society, to incessant toil, the scope of their mental energies narrowed to a few objects from the division of labour, all tending to that mechanical order and tameness incompatible with liberty; thus, perhaps, equally in danger of deteriorating and sinking into *caste* both classes yielding to the natural law of restricted adaptation to condition…

Page 3

There are several valuable varieties of apple trees of acute branch angle, which do not throw up the bark of the breeks; this either occasions the branches to split down when loaded with fruit, or if they escape this for a few years, the confined bark becomes putrid and produces canker which generally ruins the tree. We have remedied this by a little attention in assisting the rising of the bark with the knife. Nature must not be charged with the malformation of these varieties; at least had she formed them, as soon as she saw her error she would have blotted out her work.

Pages 9 and 10 (footnote)

We have never yet found one individual apple plant, raised from seed, to be the counterpart of another; but differing even in every part and habit, in bud, leaf, flower, fruit, seed, bark, wood, root; in luxuriance of growth; in hardihood; in being suited for different soils and climates, some thriving in the very moist, others only in the dry; in the disposition of the branches, erect, pendulous, horizontal; in earliness and comparative earliness of leaf, of flower of fruit.'

We hope the above remarks will not be lost on those who have the management of the sowing, planting and thinning of woods, and that they will always have selection in view. Although numerous varieties are derived from the seed of one tree, yet if that tree be of a good breed, the chances are greatly in favour of this progeny being also good.

Page 67

Our common larch like almost every other kind of tree consists of numberless varieties, which differ considerably in quickness of growth, ultimate size, and value of timber. This subject has been much neglected. We are, however, on the eve of great improvements in arboriculture; the qualities and habits of varieties are just beginning to be studied. It is also found that the uniformity in each kind of wild growing plants called *species* may be broken down by art or culture and that when once a breach is made, there is almost no limit to disorder, the mele that ensues being nearly incapable of reduction.

Page 76:

The consequences are now being developed of our deplorable ignorance of, or inattention to, one of the most evident traits of natural history, that vegetables as well as animals are generally liable to an almost unlimited diversification, regulated by climate, soil, nourishment, and new commixture of already formed varieties. In those with which man is most intimate, and where his agency in throwing them from their natural locality and dispositions has brought out this power of diversification in stronger shades, it has been forced upon his notice, as in man himself in the dog, horse, cow, sheep, poultry.- in the apple, Pear, plum, gooseberry, potato, pea, which sport in infinite varieties, differing considerably in size, colour, taste, firmness of texture, period of growth, almost in every recognisable quality. In all these kinds man is influential in preventing deterioration, by careful selection of the largest or most valuable as breeders; but in timber trees the opposite course has been pursued. The large growing varieties being so long of coming to produce seed, that many plantations are cut down before they reach this maturity, the small growing and weakly varieties, known by early and extreme seeding, have been continually selected as reproductive stock, from the ease and conveniency with which their seed could be procured; and the husks of several kinds of these invariably kiln dried, in order that the seeds might be the more easily extracted! May we then wonder that our plantations are occupied by a sickly short lived puny race, incapable of supporting existence in situations where their own kind had formerly flourished - particularly evinced in the genus Pinus more particularly in the species Scots fir; so much inferior to those of Nature's own rearing, where only the stronger, more hardy soil, suited varieties can struggle forward to maturity and reproduction?

We say that the rural economist should pay as much regard to the breed or particular variety of his forest trees, as he does to that of his live stock of horses, cows, and sheep. That nurserymen should attest the variety of their timber plants, sowing no seeds but those gathered from the largest, most healthy, and luxuriant growing trees, abstaining from the seed of the prematurely productive, and also from that of the very aged and over mature; as they, from animal analogy, may be expected to give an infirm progeny, subject to premature decay.

Pages 106-108

When woods are planted of various kinds of timber, the stronger, larger growing kinds will sometimes acquire room by overwhelming the smaller: but when the forest is of one kind of tree, and too close, all suffer nearly alike, and follow each other fast in decay, as their various strength of constitution gives way; unless, from some negligence or defect in planting, a portion of the plants have come away quickly, and the others hung back sickly for several years, so that the former might master the latter: or when some strong growing variety overtops its congeners. In the natural forest of America, when a clearance by any means is effected, the young seedlings, generally all of one kind, spring up so numerous, that, choaking each other, they all die together in a few years. This close springing up and dying is sometimes repeated several times over; different kinds of trees rising in succession, till the seeds in the soil be so reduced as to throw up plants so far asunder as to afford better opportunity for the larger growing varieties to develop their strength; and, overpowering the less, thus acquire spread of branches commensurate to the height, and thence strength of constitution sufficient to bear them forward to large trees.

Pages 153-154

Indeed the difference of quality in timber depends chiefly on the infinite varieties existing in what is called Species, though soil and climate have no doubt considerable influence, both in forming the variety, and in modifying it while growing. Of varieties those which have the thinnest bark under equal exposure have the hardest wood.

Page 202

In like manner, in all the other relations, we see Nature especially accommodating the character of each individual plant, to the exigencies of its particular situation. In the interior of woods, the wind can exert a far less mechanical effect on individual trees; and therefore, while they axe positively determined to push upwards towards the light, they are negatively permitted to do, so by the removal of any necessity to thicken their trunks, for the sake of greater strength, and to contract the height of them, in order to afford the blast a shorter lever against the roots. But, with trees in an open situation, all this is widely different. There they are freely exposed to the wind, and the large expansion of their branches, gives every advantage to the violence of the storm. Nature accordingly, bestows greater proportional thickness, and less proportional elevation on trees, which are isolated, or nearly so; while their system of root, which, by necessity, is correlatively proportional to their system of top, affords likewise heavier ballast, and a stronger anchorage, in order to counteract the greater spread of sail, displayed in the wider expansion of the branches. Every individual tree is thus a beautiful system of qualities specially relative to the place which it holds in creation of provisions admirably accommodated to the peculiar circumstances of its case.

Pages 261- 263

Gardeners certainly experience the branches and roots of crab apple to be harder than the varieties with thicker bark, larger more downy leaves, and larger fruit. The largest growing apple varieties, however, are not the above mentioned mild varieties, but those which have a pretty close approximation to the crab. We have taken slips from some of the very largest of our pear trees, and having placed them close to the ground on young stocks, have found they threw out spines and rectangular branching similar to crabs. Those most dissimilar to the crab have thick annual shoots, without any lateral rectangular branching, and very thick bark; they have been gradually bred to this condition by repeated sowing, always choosing the seed of those partaking most of these qualities for resowing, their disposition to vary to mildness being at the same time influenced in some measure by culture and abundant moist nourishment: but these mild varieties; although they throw out a strong annual shoot while young, seldom or never reach to any considerable size of tree, unless they are nourished by crab roots, their own roots being soft and fleshy, and incapable of foraging at much depth or distance. Their branches and twigs as they get old are also very soft and friable, covered with a thick bark, but the timber of the stem is very little inferior in hardness to crab timber.

We ask if even the fact of these unnaturally tender varieties (obtained by long continued selection, probably assisted by culture, soil and climate, and which, without the cherishing of man, would soon disappear), being of rather more porous texture of wood goes any length to prove our author's assertion? We have paid some attention to the fibre of the genus Pyrus, and find that the Siberian crabs have by far the smallest vessels. Having grafted the large Fulwood upon the smallest Red Siberian Crab, or Cherry-apple, the new wood layers above the junction swelled to triple the thickness of those below. By ingrafting other kinds upon other stocks we have found the reverse to take place n[o] doubt owing to those with largest vessels swelling the most, there being the same number of vessels above and below the junction, each corresponding, or being a continuation of the other. But this small Siberian crab, when ingrafted upon a common crab, grew fully as quickly during several years as the Fulwood under the same circumstances; and the timber though of much finer texture, scarcely exceeded the other in hardness. Sir Henry tells us, that the oak is less durable in Italy and Spain than in England. We tell Sir Henry, that the redwood pitch pine from Georgia and the Floridas, on the confines of the torrid zone, is more durable than the red wood pine from Archangel, on the confines of the frigid zone. But does this fact regarding the oak of the south of Europe prove any thing regarding the oak of England,- that it will always be deteriorated by culture for several years after planting, or that the quality may not suffer as much from slowness of growth as from fastness, or from the climate being too cold as from being too warm?

[**Matthew's own footnote to page 285**]: *The fineness of vessel or fibre of the Siberian crab may be induced by the arid warm air the continued radiation of heat and light upon the portion above ground and the coldness of the ground around the roots during the short summer in Siberia where the air and surface of the ground is warm and vegetation progressive while the ground remains frozen at a small depth Like all varieties of plants habituated to colder climate the Siberian crab developes its leaves under less heat than varieties of the same kind which have been habituated to milder We have not taken Sir Henry in the literal sense Timber is well known to decay sooner in a warm than in a cold country'.*

The reason why Highland Scots oak spokes are superior to English is because the latter are generally split from out the refuse of the timber cut for naval purposes principally the branches and tops of large trees whereas those from the Highlands of Scotland are from the root cuts of copse. We believe most carpenters of Scotland are aware of this. The oak from the Highlands of Scotland is however for the most part of excellent quality growing generally on dry gravel and rock not on cold moist clayey soils. The hardest we have ever seen was from a steep dry gravel bank of south exposure, in an open situation, much exposed to the western breeze. The Highland oak from these soils is generally of a greyish colour, and very dense; whereas that from moist soils is often reddish brown, and defective. Should Sir Henry weigh portions of oak from these soils in a pair of material, in place of mental scales, we think his conclusions would be somewhat different. The strongest hardest ash we have seen, was cut from a hard, dry, adhesive clay, of course a young tree.

Sir Henry, speaking of the Western Highlands and Islands of Scotland, states that "it is from a want of soil, and not of climate, that woods of any given extent cannot be got up in these unsheltered but romantic situations." Of many situations of these bleak districts, this must be admitted, but we cannot receive it as a general fact; and even where it holds true, the want of (proper) soil, or formation of peat is a *consequence* of the want of climate, although *this* may have reacted to increase the evil. There must have been a greater warmth of climate, at least in summer when the forests grew, which lie buried in the mosses of the northern part of Scotland, and of the Orkney and Shetland Islands, as some kinds of timber are found in situations, where such kinds by no circumstances of gradual shelter under the present climate could have grown. There are several indications of a greater warmth having been general throughout Britain, and even farther eastward, and that a slight refrigeration is still in progress. We instance the once numerous vineyards of England,- the vestiges of aration so numerous upon many of our hills, where it would now be considered fruitless to attempt raising grain…

Pages 283-286

In tall trees this greater deposition on the stem, in proportion to that on the roots, twigs, and leaves, some will think instinctive; some will refer it to an effort of nature to supply the necessary strength to enable the stem to resist the great strain of the winds upon the elevated top. If it take place to a greater extent than what arises from the greater elongation of the necessary vessels of communication, perhaps it is owing to the evaporation or stagnation of the sap on the tall exposed stem, and to the considerable motion or waving of the stem by wind promoting deposition, evincing one of the deep balancings of material cause and effect, or circumstantial regulation, which mocks the wisdom of the wise.

Page 301

Our author's next implied assumption, that a tree produces best timber in a soil and climate *natural* to it (we suppose by this is meant the soil and climate where the kind of tree is naturally found growing), is, we think, at least exceedingly hypothetical; and, judging from our facts, incorrect. The natural soil and climate of a tree, is often very far from being the soil and climate most suited to its growth, *and is only the situation where it has greater power of occupancy than any other plant whose germ is present*. The pines do not cover the pine barrens of America, because they prefer such soil, or grow most luxuriant in such soil; they would thrive much better, that is, grow faster in the natural allotment of the oak and the walnut, *and also mature to a better wood in this deeper richer soil*. But the oak and the walnut banish them to inferior soil from greater power of occupancy in good soil, as the pines, in their turn, banish other plants from inferior sands - some to still more sterile location, by the same means of greater powers of occupancy in these sands. One cause considerably affecting the natural location of certain kinds of plants is, that only certain soils are suited to the preservation of certain seeds, throughout the winter or wet season. Thus many plants, different from those which naturally occupy the soil, would feel themselves at home, and would beat off intruders, were they once seated. We have had indubitable proof in this country, that Scots fir grown upon good deep loam, and strong till (what our author would call the natural soil of the oak), *is of much better quality, and more resinous, than fir grown on poor sand* (what he would call the natural soil of the Scots fir), although of more rapid growth on the loam than on the sand; and the best Scots fir we have ever seen, of equal age and quickness of growth, is growing upon Carse land (clayey alluvium).

Pages 302-303

Man's interference is useful in removing competitors, in giving it lateral room for extension, in *training* it skillfully to one leader and subordinate equality of feeders, should transplanting, early pruning up, or other cause, destroy the natural regular pyramidal disposition - not in pruning it up, thus reducing it to narrower compass, and destroying its balance to the locality.

The use of the infinite seedling varieties in the families of plants, even in those in a state of nature, differing in luxuriance of growth and local adaptation, seems to be to give one individual (the strongest best circumstance-suited) superiority over others of its kind around, that it may, by overtopping and smothering them, procure room for full extension, and thus affording, at the same time, a continual selection of the strongest, best circumstance-suited, for reproduction. Man's interference, by preventing this natural process of selection among plants, independent of the wider range of circumstances to which he introduces them, has increased the difference in varieties, particularly in the more domesticated kinds; and even in man himself, the greater uniformity, and more general vigour among savage tribes, is referrible to nearly similar selecting law - the weaker individual sinking under the ill treatment of the stronger, or under the common hardship.

As our author's premises thus appear neither self evident, nor supported by facts, it might seem unfair, at least it would be superfluous, to proceed to the consideration of his conclusions and corollaries.

Page 308

There is a deposition from the atmosphere of saline matter going on at the surface of the earth, either evaporated from the ocean, and falling with the rain and dews, or formed by gaseous combinations - most probably both. In countries where the quantity of rain is insufficient to wash this saline accumulation away into the ocean as fast as it is formed, it increases to such a degree as almost to prevent vegetation only a few of what are termed saline plants appearing. This saline accumulation in warm dry countries bears considerable analogy to tannin deposit in cold countries.

Page 325 {Footnote}

Sea salt, perhaps also nitre and other salts, will be serviceable in a moist country, or far from the sea, where the plants and water contain little saline matter, and probably pernicious in a dry climate, where the plants and water generally contain much saline matter.

Page 325

And besides, we have found varieties of the same kind or species of tree *some of them adapted to prosper in dry air and soil, and others in moist air and soil.* Although the above causes prevent a positive limitation of certain kinds of trees to certain soils, yet there are some which have superior adaptation to moist soils and others to dry; some whose roots from their fibrous soft character, can only spread luxuriantly on light, soft, or mossy soils, and others, whose roots have power to permeate the stiffest and most obdurate. The above explanations will account for much of the incongruity which we find in authors regarding the adaptation of certain kinds of timber to certain soils.

Page 335

The highest latitude to which a tree, or any other kind of plant, reproducing by see, naturally extends, depending on the ripening of the seed, and also on the power of occupancy, is however different from that where it will grow, when ripe seeds are procured from the coldest place where they ripen, and all the competitors removed; and under the system of shelter belts, hardy pine nurses, and seeds from the nearest place where they ripen, we have no doubt that oaks may be extended to a colder situation than Nature herself would have placed them in. For the higher more bleak portion of the country, we would recommend acorns grown in Scotland, in preference to those imported from England. We have several times observed wheat, the seed of which had been imported from England, sustain blight and other injuries in a cold moist autumn when a portion of the same field, sown of Scots seed, at the same time as the other, and under the very same circumstances, was entirely free from injury.

Pages 357-358

== *NTA* APPENDIX BEGINS ==

Matthew's Note B

There is a law universal in nature tending, to render every reproductive being the best possibly suited to its condition that its kind, or that organized matter, is susceptible of, which appears intended to model the physical and mental or instinctive powers, to their highest perfection, and to continue them so. This law sustains the lion in his strength the hare in her swiftness and the fox in his wiles. As Nature, in all her modifications of life, has a power of increase far beyond what is needed to supply the place of what falls by Time's decay, those individuals who possess not the requisite strength, swiftness, hardihood, or cunning, fall prematurely without reproducing -either a prey to their natural devourers, or sinking under disease, generally induced by want of nourishment, their place being occupied by the more perfect of their own kind, who are pressing on the means of subsistence. The law of entail, necessary to hereditary nobility, is an outrage on this law of nature which she will not pass unavenged - a law which has the most debasing influence upon the energies of a people, and will sooner or later lead to general subversion, more especially when the executive of a country remains for a considerable time efficient, and no effort is needed on the part of the nobility to protect their own, or no war to draw forth or preserve their powers by exertion. It is all very well, when in stormy times, the baron has every faculty trained to its utmost ability in keeping his proud crest aloft. How far hereditary nobility, under effective government, has operated to "retard the march of intellect," and deteriorate the species in modern Europe, is an interesting and important question. We have seen it play its part in France; we see exhibition of its influence throughout the Iberian peninsula, to the utmost degradation of its victims. It has rendered the Italian peninsula, with its islands, a blank in the political map of Europe. Let the panegyrists of hereditary nobility, primogeniture, and entail, say what these countries might not have been but for the baneful influence of this unnatural custom. It is an eastern proverb, that no king is many removes from a shepherd. Most conquerors and founders of dynasties have followed the plough or the flock. Nobility, to be in the highest perfection, like the finer varieties of fruits, independent of having its vigour excited by regular married alliance with wilder stocks, would require stated complete renovation, by selection anew from among the purest crab. In some places, this renovation would not be so soon requisite as in others, and judging from facts, we would instance Britain as perhaps the soil where nobility will continue the longest untainted. As we advance nearer to the equator, renovation becomes sooner necessary, excepting at high elevation - in many places, every third generation, at least with the Caucasian breed, although the finest stocks be regularly imported. This renovation is required as well physically as morally.

It is chiefly in regard to the interval of time between the period of necessary feudal authority, and that when the body of the population having acquired the power of self-government from the spread of knowledge, claim a community of rights, that we have adverted to the use of war. The manufacturer, the merchant, the sailor, the capitalist, whose mind is not corrupted by the indolence induced under the law of entail, are too much occupied to require any stimulant beyond what the game in the wide field of commercial adventure affords. A great change in the circumstances of man is obviously at hand.

In the first step beyond the condition of the wandering savage, while the lower classes from ignorance remained as helpless children, mankind naturally fell into clans under paternal or feudal government; but as children, when grown up to maturity, with the necessity for protection, lose the subordination to parental authority, so the great mass of the present population requiring no guidance from a particular class of feudal lords, will not continue to tolerate any hereditary claims of authority of one portion of the population over their fellow-men; nor any laws to keep up rank and wealth corresponding to this exclusive power.- It would be *wisdom* in the noblesse of Europe to abolish every claim or law which serves to point them out a separate class, and, as quickly as possible, to merge themselves into the mass of the population. It is a law manifest in nature, that when the use of any thing is past its existence is no longer kept up.

Although the necessity for the existence of feudal lords is past, yet the same does not hold in respect to a hereditary head or King; and the stability of this head of the government will, in no way, be lessened by such a change. In the present state of European society, perhaps no other rule can be so mild and efficient as that of a liberal benevolent monarch, assisted by a popular representative Parliament. The poorest man looks up to his king as his own, with affection and pride, and considers him a protector; while he only regards the antiquated feudal lord with contempt. The influence of a respected hereditary family as head of a country, is also of great utility in forming a principle of union to the different members, and in giving unity and stability to the government.

In respect to our own great landholders themselves, we would ask, where is there that unnatural parent -that miserable victim of hereditary pride - who does not desire to see his domains equally divided among his own children? The high paid sinecures in church and state will not much longer be a great motive for keeping up a powerful family head, whose influence may burthen their fellow-citizens with the younger branches. Besides, when a portion of land is so large, that the owner cannot have an individual acquaintance and associations with every stream, and bush, and rock, and knoll, the deep enjoyment which the smaller native proprietor would have in the peculiar features, is not called forth, and is lost to man. The abolition of the law of entail and primogeniture, will, in the present state of civilization, not only add to the happiness of the proprietor, heighten morality, and give much greater stability to the social order, but will also give a general stimulus to industry and improvement, increasing the comforts and elevating the condition of the operative class.

In the new state of things which is near at hand, the proprietor and the mercantile class will amalgamise,- employment in useful occupations will not continue to be held in scorn,- the merchant and manufacturer will no longer be barely tolerated to exist, harassed at every turn by imposts and the interference of petty tyrants;- Government, instead of forming an engine of oppression, being simplified and based on morality and justice, will become a cheap and efficient protection to person and property; and the necessary taxation being levied from property alone, every individual will purchase in the cheapest market, and sell the produce of his industry in the dearest. This period might perhaps be accelerated throughout Europe, did the merchants and capitalists only know their own strength Let them, as citizens of the world, hold annual congress in some central place, and deliberate on the interests of man, which is their own, and throw the whole of their influence to support liberal and just governments, and to repress slavery, crime, bigotry - tyranny in all shapes. A Rothschild might earn an unstained fame, as great as yet has been attained by man, by organizing such a power, and presiding at its councils.

== Note F ==

(This part of the Appendix begins with Matthew's geological observations and is then immediately followed by a concentrated body of several pages of natural selection relevant text. Interestingly, Darwin (1839) published an extremely erroneous Royal Society geological paper on the parallel roads of Glen Roy, which he discussed in detail with Robert Chambers.)

In the case of the upper carse on the Tay Firth, there is evidence both from its vestiges and from records, that it had occupied, at least, the entire firth, or sea-basin, above Broughty Ferry, and that about 50 square miles of this carse has been carried out into the German Ocean by the strong sea tide current a consequence of the lowering of the German Ocean and of the deepening of the outlet of this sea basin at Broughty Ferry, apparently by this very rapid sea-tide current. This carse appears to have been a general deposition at the bottom of a lake having only a narrow outlet communicating with the sea, and probably did not rise much higher than the height of the bottom of the outlet at that time.

An increase of deposition of alluvium, or prevention of decrease may, in many cases, be accomplished by artificial means. The diminution of the carse of the Tay was in rapid progress about sixty years ago, the sea-bank being undermined by the waves of the basin, the clay tumbling down, becoming diffused in the water, and being carried out to sea, by every ebbing tide, purer water returning from the ocean the next tide- flow. This decrease was stopped by the adoption of stone embanking and dikes. A small extension of the carses of present high-water level in the upper part of the firths of Tay and Forth, has lately been effected, by forming brushwood stone and mud dikes, to promote the accumulation.

In doing this, the whole art consists in placing obstructions to the current and waves, so that whatever deposition takes place at high water or at the beginning of the flood- tide, when the water is nearly still, may not again be raised and carried off.

Notwithstanding this accumulation, and also the prevention of further waste of the superior carse, the deepening of the Tay Firth formerly carse, and of the gorge at Broughty Ferry, seems still in progress, and could not, without very considerable labour, be prevented In the case however, of the sea basin of Montrose, a little labour, from the narrowness of the gorges, would put it in a condition to become gradually filled with mud. Not a great deal more expenditure than what has sufficed to erect the suspension bridge over its largest outlet, would have entirely filled up this outlet, and the smaller outlet might have been also filled to within several feet of high-water, and made of sufficient breadth only, to emit the water of the river which flows into the basin. The floated sand and mud of this river, thus prevented from being carried out to sea, would in the course of years, completely fill up the basin.

From some vestiges of the upper carse, as well as of the lower or submarine carse, in situations where their formation cannot easily be traced to any local cause, it seems not improbable that the basin of the German sea itself, nearly as far north as the extent of Scotland, had at one time been occupied with a carse or delta, a continuation of Holland, formed by the accumulation of the diluvium of the rivers which flow into this basin, together with the molluscous exuviae of the North Sea, and the abrasion of the Norwegian coast and Scottish islands, borne downward by the heavy North Sea swell.

In the case of the delta of Holland having extended so far northward, a subsidence of the land or rising of the sea, so as to form a passage for the waters round Britain, must have occurred. The derangement at several places, of the fine wavy stratification of these carses, and the confusedly heaped-up beds of broken sea-shells, shew that some great rush of water had taken place, probably when Belgium was dissevered from England. Since the opening of the bottom of the gulf, the accumulation may have been undergoing a gradual reduction, by more diffused mud being carried off from the German Sea into the Atlantic and North Sea, than what the former is receiving the same process taking place here as has been occurring in the basin of the Tay. The large sandbanks on the Dutch and English coast,- in some places, such as the Goodwin Sands, certainly the heavier, less diffusible part of the former alluvial country, and portions of these alluvial districts being retained by artificial means,- bear a striking resemblance to the sand banks of the sea basin of the Tay - the less diffusible remains of the removed portion of the alluvium which had once occupied all that basin, and to the remaining portion of the alluvium also retained by artificial means.

Here Matthew's discovery of Natural Selection continues

Throughout this volume, we have felt considerable inconvenience, from the adopted dogmatical classification of plants, and have all along been floundering between species and variety, which certainly under culture soften into each other. A particular conformity, each after its own kind, when in a state of nature, termed species, no doubt exists to a considerable degree. This conformity has existed during the last forty centuries. Geologists discover a like particular conformity - fossil species - through the deep deposition of each great epoch, but they also discover an almost complete difference to exist between the species or stamp of life, of one epoch from that of every other. We are therefore led to admit either of a repeated miraculous creation; or of a power of change, under a change of circumstances, to belong to living organized matter, or rather to the congeries of inferior life, which appears to form superior. The derangements and changes in organized existence, induced by a change of circumstance from the interference of man, affording us proof of the plastic quality of superior life, and the likelihood that circumstances have been very different in the different epochs, though steady in each tend strongly to heighten the probability of the latter theory.

When we view the immense calcareous and bituminous formations, principally from the waters and atmosphere, and consider the oxidations and depositions which have taken place, either gradually, or during some of the great convulsions, it appears at least probable, that the liquid elements containing life have varied considerably at different times in composition and in weight; that our atmosphere has contained a much greater proportion of carbonic acid or oxygen; and our waters, aided by excess of carbonic acid, and greater heat resulting from greater density of atmosphere, have contained a greater quantity of lime and other mineral solutions. Is the inference then unphilosophic that living things which are proved to have a circumstance-suiting power a very slight change of circumstance by culture inducing a corresponding change of character - may have gradually accommodated themselves to the variations of the elements containing them, and, without new creation, have presented the diverging changeable phenomena of past and present organized existence.

The destructive liquid currents, before which the hardest mountains have been swept and comminuted into gravel, sand, and mud, which intervened between and divided these epochs, probably extending over the whole surface of the globe, and destroying nearly all living things, must have reduced existence so much, that an unoccupied field would be formed for new diverging ramifications of life, which from the connected sexual system of vegetables, and the natural instincts of animals to herd and combine with their own kind, would fall into specific groups, these remnants, in the course of time moulding and accommodating their being anew to the change of circumstances, and to every possible means of subsistence, and the millions of ages of regularity which appear to have followed between the epochs, probably after this accommodation was completed affording fossil deposit of regular specific character.

There are only two probable ways of change - the above, and the still wider deviation from present occurrence.- of indestructible or molecular life (which seems to resolve itself into powers of attraction and repulsion under mathematical figure and regulation, bearing a slight systematic similitude to the great aggregations of matter), gradually uniting and developing itself into new circumstance suited living aggregates, without the presence of any mould or germ of former aggregates, but this scarcely differs from new creation, only it forms a portion of a continued scheme or system.

In endeavouring to trace in the former way, the principle of these changes of fashion which have taken place in the domiciles of life, the following questions occur: Do they arise from admixture of species nearly allied producing intermediate species? Are they the *diverging ramifications* of the living principle under modification of circumstance? Or have they resulted from the combined agency of both? Is there only one living principle? Does organized existence, and perhaps all material existence consist of one Proteus principle of life capable of gradual circumstance-suited modifications and aggregations without bound under the solvent or motion giving principle, heat or light? There is more beauty and unity of design in this continual balancing of life to circumstance, and greater conformity to those dispositions of nature which are manifest to us, than in total destruction and new creation. It is improbable that much of this diversification is owing to commixture of species nearly allied all change by this appears very limited, and confined within the bounds of what is called Species: the progeny of the same parents, under great difference of circumstance, might, in several generations, even become distinct species incapable of co reproduction.

The self-regulating adaptive disposition of organized life may, in part, be traced to the extreme fecundity of Nature, who, as before stated, has in all the varieties of her offspring, a prolific power much beyond (in many cases a thousandfold) what is necessary to fill up the vacancies caused by senile decay. As the field of existence is limited and pre-occupied, it is only the hardier, more robust, better suited to circumstance individuals, who are able to struggle forward to maturity, these inhabiting only the situations to which they have superior adaptation and greater power of occupancy than any other land the weaker less circumstance-suited being prematurely destroyed. This principle is in constant action, it regulates the colour, the figure, the capacities, and instincts; those individuals of each species, whose colour and covering are best suited to concealment or protection from enemies, or defence from vicissitude and inclemencies of climate, whose figure is best accommodated to health, strength, defence, and support; whose capacities and instincts can best regulate the physical energies to self-advantage according to circumstances - in such immense waste of primary and youthful life, *those* only come forward to maturity from the strict ordeal by which Nature tests their adaptation to her standard of perfection and fitness to continue their kind by reproduction.

From the unremitting operation of this law acting in concert with the tendency which the progeny have to take the more particular qualities of the parents, together with the connected sexual system in vegetables, and instinctive limitation to its own kind in animals, a considerable uniformity of figure, colour, and character, is induced, constituting species; the breed gradually acquiring the very best possible adaptation of these to its condition which it is susceptible of, and when alteration of circumstance occurs, thus changing in character to suit these as far as its nature is susceptible of change.

This circumstance-adaptive law, operating upon the slight but continued natural disposition to sport in the progeny (seedling variety), does not preclude the supposed influence which volition or sensation may have over the configuration of the body. To examine into the disposition to sport in the progeny, even when there is only one parent, as in many vegetables, and to investigate how much variation is modified by the mind or nervous sensation of the parents, or of the living thing itself during its progress to maturity; how far it depends upon external circumstance and how far on the will irritability and muscular exertion is open to examination and experiment. In the first place, we ought to investigate its dependency upon the preceding links of the particular chain of life, variety being often merely types or approximations of former parentage; thence the variation of the family, as well as of the individual, must be embraced by our experiments.

This continuation of family type, not broken by casual particular aberration, is mental as well as corporeal, and is exemplified in many of the dispositions or instincts of particular races of men. These innate or continuous ideas or habits, seem proportionally greater in the insect tribes, those especially of shorter revolution; and forming an abiding memory, may resolve much of the enigma of instinct, and the foreknowledge which these tribes have of what is necessary to completing their round of life, reducing this to knowledge, or impressions, and habits, acquired by a long experience. This greater continuity of existence, or rather continuity of perceptions and [i]mpressions, in insects, is highly probable; it is even difficult in some to ascertain the particular stops when each individuality commences, under the different phases of egg larva pupa or if much consciousness of individuality exists. The continuation of reproduction for several generations by the females alone in some of these tribes, tends to the probability of the greater continuity of existence, and the subdivisions of life by cuttings, at any rate must stagger the advocate of individuality.

Among the millions of *specific varieties* of living things which occupy the humid portion of the surface of our planet, as far back as can be traced, there does not appear, with the exception of man, to have been any particular engrossing race, but a pretty fair balance of powers of occupancy,- or rather, most wonderful variation of circumstance parallel to the nature of every species, as if circumstance and species had grown up together. There are indeed several races which have threatened ascendency in some particular regions, but it is man alone from whom any general imminent danger to the existence of his brethren is to be dreaded. As far back as history reaches, man had already had considerable influence, and had made encroachments upon his fellow denizens, probably occasioning the destruction of many species, and the production and continuation of a number of varieties or even species, which he found more suited to supply his wants, but which, from the infirmity of their condition - not having undergone selection by the law of nature, of which we have spoken cannot maintain their ground without his culture and protection. It is however only in the present age that man has begun to reap the fruits of his tedious education, and has proven how much "knowledge is power." He has now acquired a dominion over the material world, and a consequent power of increase, so as to render it probable that the whole surface of the earth may soon be overrun by this engrossing anomaly, to the annihilation of every wonderful and beautiful variety of animated existence, which does not administer to his wants principally as laboratories of preparation to befit cruder elemental matter for assimilation by his organs.

== END==

ABOUT THE AUTHOR

Dr Mike Sutton is an international award-winning criminologist. He is co-founder and Editor in Chief of the Internet Journal of Criminology. One of his interests is the importance of debunking long entrenched academic myths.

For the academic year 1998 to 1999, Mike was co-recipient with David Mann of the British Journal of Criminology Prize, awarded for their virtual ethnography article on high tech crime and hackers that: *'Most significantly contributed to the knowledge and understanding of criminology and criminal justice issues.'* Mike has written many articles, book chapters and government reports on a variety of criminological issues, including evaluating the £50 million Safer Cities national crime reduction program, various attitude change programs and the use of the Internet by Far Right hate groups. He has spoken on his original criminological research at academic conferences, police and other crime reduction practitioner and policy meetings in the USA, Canada, Brazil, the Netherlands, Germany, Poland, Italy, England Scotland, Northern Ireland, Southern Ireland and Wales In the 1990's, Mike originated the Market Reduction Approach to the theft.

In 2013, Sutton used innovative high technology research methods on the Internet to make the discovery that proves Charles Darwin and the world's leading experts on evolution by natural selection were wrong to claim that no naturalist, no biologist, or no one whatsoever read Patrick Matthew's (1831) prior published discovery of macroevolution by natural selection before Darwin and Wallace replicated it in 1858. Sutton originally discovered the fact that three out of 24 people he found did cite Matthew's book before 1858 were respectively: prominent naturalist geologists, ornithologists, botanist biologists at the epicentre of pre-1858 influence on Darwin and Wallace.

Dr Mike Sutton.

(Photograph by Andy Sutton, who is, by the way, not a relative of the author.)

INDEX

OBSERVATIONS ABOUT THIS BOOK

'Revealing the results of ground breaking Big Data research, "Nullius in Verba" draws back the curtain to expose the shameless role of the Royal Society in hiding facts and perpetuating science myths in a purposefully orchestrated passing over and trivialization of one of the greatest minds in the history of Scotland. The scientific establishment disingenuously diminished Patrick Matthew, a noted forester, horticulturist and naturalist. Influential scientists, including Charles Darwin sought to bury in oblivion the first, precise and more accurate published work on evolution by Natural Selection, which was pillaged and plundered by Darwin and several of his associates. Criminologist Dr Mike Sutton has uncovered a can of wriggling worms. Following his original discoveries, the serious matter regarding why the Royal Society allowed and rewarded such dishonorable activity, which breached its famous acceptance of the Arago Case ruling for deciding both first and foremost priority for original scientific discovery, cries out for further enquiry.'

Howard L. Minnick
Major, Corps of Engineers
United States Army (Ret.)
Botanist, Range Conservationist
& 3rd Great Grandson of Patrick Matthew

www.ingramcontent.com/pod-product-compliance
Lightning Source LLC
Chambersburg PA
CBHW061436180526
45170CB00004B/1437